HARVARD ECONOMIC STUDIES

VOLUME XCVIII

SCHEDULING OF PETROLEUM REFINERY OPERATIONS

by
Alan S. Manne

HARVARD UNIVERSITY PRESS

Cambridge, Massachusetts

1963

Second Printing

Library of Congress Catalog Card Number 56-6518

Printed in the United States of America

Preface

This study grew out of a doctoral dissertation carried out under the direction of Dean Edward Mason and Professor Wassily Leontief of Harvard University. Shortly thereafter, Professor J. K. Galbraith suggested the desirability of further research into the problems of refinery management. I owe a particular debt of gratitude to all three of my former instructors.

More recently, at the encouragement of Dr. Charles Hitch, this work has been carried on at The RAND Corporation. To acknowledge my indebtedness to others at RAND would surely require a long list, but two colleagues are to be singled out for having provided a constant source of stimulation: Dr. George Dantzig and Dr. Harry Markowitz.

And finally, thanks are due to my wife, Jacqueline, who was willing to combine a courtship with the proof-reading of these pages.

Alan S. Manne

Santa Monica, California
July, 1954

Contents

I. Introduction   1

Preliminary observations. A summary. Orientation of the book

Appendix: A Glossary of Technical Terms

II. Conventional Methods of Refinery Economic Analysis   10

Standard cost accounting formulae. Gasoline replacement value estimates. Comparison of multiple alternatives

III. A Crude Oil Allocation Problem   21

Introduction. Existing methods of crude oil evaluation. A mathematical analysis of the scheduling problem. Computational procedure. Summary and conclusions

IV. A Naphtha Reforming Problem   45

Introduction. Basic economic assumptions. Basic engineering assumptions. Product yields and input costs. The graphical solution. Algebraic procedure. Summary and conclusions

V. A Gasoline Blending Problem 70

Basic economic assumptions. The blending technology. The mathematical model. The numerical analysis. Results of the numerical analysis. Summary and conclusions

VI. Cracking, Recycling, and Blending — An Integrated Refinery Problem 109

A preview. Operating conditions and product yields for a simplified refinery. Economic and technological assumptions. The mathematical model. The numerical analysis, run no. 1. Run no. 2, variations in the price of Number 6 fuel oil. Run no. 3, inter-product flexibility between gasoline and Number 1 fuel oil. Run no. 4, a new investment problem. Summary and conclusions

Appendix: Notes on Parametric Linear Programming 172

VII. The Economist and the Operations Scheduler 178

What can the economist currently contribute to the refinery programming problem? What is the economist currently unable to contribute to the refinery programming problem? What can the economist learn about the refinery operator? How can the economist make better estimates of sector-wide and economy-wide capabilities?

Tables

II.1. Summary of Economics — Processing 10^4 Barrels per Calendar Day (B/CD) of East Texas Crude   17

III.1. Hypothetical Refined Product Yields, Realizations, and Expenses on the ith Crude Oil at the jth Refinery   24

III.2. Hypothetical Company-wide Crude Oil Availabilities and Distillation Capacities   29

III.3. Hypothetical Net Realization upon Crude Oil i at Refinery j, p_{ij} 29

III.4. Schedule of x_{ij}, Crude Oil Allocations   31

III.5. Schedule of x_{ij}. First Phase in Determining a Basic Feasible Solution   35

III.6. Schedule of x_{ij}. Second Phase in Determining a Basic Feasible Solution   37

III.7. Schedule of x_{ij}. Initial Basic Feasible Solution   37

III.8. Net Company Realizations, p_{ij}, and Dual Variables, u_i and v_j. Initial Basic Feasible Solution   38

III.9. Dual Variables u_i and v_j; Activity Surpluses $p_{ij} - u_i - v_j$. Initial Basic Feasible Solution   39

III.10. Modifications Required in Initial x_{ij} Shipping Schedule in Order to Introduce x_{33} 40

III.11. Shipping Schedules, x_{ij}, for Seven Successive Base Shifts, Rounded Off   42

III.12. Dual Variables, u_i and v_j; Activity Surpluses $p_{ij} - u_i - v_j$. Optimum Solution, Seventh Basis — 43

IV.1. Single Pass Naphtha Reforming Costs — 48

IV.2. Ethyl Fluid Consumption as a Function of Reformer Charge and Gasoline Yield — 64

IV.3. Comparison of Two Methods of Predicting TEL Consumption — 66

IV.4. Comparison of Optimal Reforming Conditions — Nomogram and Algebraic Methods — 68

V.1. Gasoline Blending Problem — Definition of Various Quantities — 84

V.2. Comparison of Partial Derivatives for $t = 60$, Run No. 1 — 97

V.3. Results of Computations — 99

VI.1. Cuts from $38.0°$ API Gravity Crude Oil — 113

VI.2. Product Streams, before Blending — 114

VI.3. Yields of Conversion Products, Single-Pass Cracking — 115

VI.4. Product Streams, after Blending — 116

VI.5. Coefficients for Programming a Simplified Thermal Cracking Refinery — 129

VI.6. Five Ethyl Fluid Concentration Levels and Corresponding Values for k_2, p_{21}, and p_{26} — 139

VI.7. Calculation of $p_1, p_2, \ldots, p_7$ Coefficients — 144

VI.8. Product-Mix Summary, Run No. 1 — 147

VI.9. Solution to Run No. 1, Basic Case, $k_1 = 3.0$ ml TEL per Gallon of Regular Grade Gasoline — 149

VI.10. Comparative Product-Mix Summary, Runs Nos. 1 and 2 — 153

VI.11. Activity Levels, Optimal Bases, Runs Nos. 1 and 2 — 154

VI.12. Product-Mix Summary, Run No. 3; Substitution between Cracked Gasoline and No. 1 Fuel Oil — 157

VI.13. Derivation of Capital Costs for Run No. 4 — 161

VI.14. Changes in Matrix, Table VI.5, for the "New Investment" Problem — 162

VI.15. Optimum Product-Mix and Construction Program for the "New Investment" Problem — 165

VI.16. Solution to Run No. 4, New Investment Problem — 166

VI.17. Product-Mix, Charge Capacities, and Payoff at Five Levels of Capital Investment — 168

Figures

1. Simplified Flow Diagram for Thermal and Vacuum Flashing Schemes   18

2. A Schematic Diagram for Thermal Reforming and Ethyl Fluid Blending   46

3. Processing Costs and Initial Investment Costs vs. Reformer Unit Size. Single Pass Reforming of Virgin Heavy Naphthas   49

4. Summary of Nomograms, Naphtha Reforming Problem   55

4g. (No Title)   56

5. Optimum Fuel Yield vs. Price of Fuel Oil   58

6. TEL Consumption vs. Price of Fuel Oil   59

7. Correlation between Gasoline Yield, Octane Number of Reformer Charge Stock, and Octane Number of Gasoline Product   61

8. TEL Concentration Requirements for 70 Octane Finished Gasoline   62

9. Ethyl Fluid Consumption, E, as a Function of Reformer Charge, X, and Yield, Y 65

10. Union Oil Co. Gasoline Blending Problem   74

11. Octane Number vs. Percent of Thermally Cracked Component in Binary Blend. Three TEL Levels   78

12a. Cutting Temperature vs. x_{31} 79

12b. Clear and Leaded Octane Numbers vs. x_{31} 80

12c. Percent of Sulfur Content and Volatility Index vs. x_{31} 81

13a. Run No. 1, Payoff and Total Gasoline Output   93

13b. Run No. 1, Gasoline Components x_{42} and x_{45} 93

13c. Run No. 1, TEL Concentration Levels x_{21} and x_{22} 94

13d. Run No. 1, Lagrangean Multipliers, u_5 and u_6 94

13e. Run No. 1, Relative Octane Numbers, $f_5(x) \div z_1$, and $f_6(x) \div z_2$ 95

14a. Total Payoff, $g(x)$, Four Cases   98

14b. Total Output of 7600 Gasoline, Four Cases   100

14c. Total Output of 76 Gasoline, Four Cases   100

14d. Total Consumption of TEL, Four Cases   101

15. Schematic Diagram of Thermal Cracking Refinery   111

16. (No Title)   119

17. Properties of Component Materials for No. 1 Fuel Oil   141

18. Run No. 3. Three "Efficient Points"   158

19. Payoff-Investment Curve, Run No. 4   169

Scheduling of Petroleum Refinery Operations

Chapter I

Introduction

1. PRELIMINARY OBSERVATIONS

The reader of this volume is likely to be one of two distinct types — either a general-purpose economist or else a specialist in petroleum engineering. The book has been written in the belief that each of these groups has something to contribute to the other, and that the barrier of professional jargon is not an insuperable one. The aim has been to report on several examples of the actual use of mathematical economics in representative oil refinery scheduling problems. This book is by no means intended as an exhaustive treatment of the subject. On the contrary, there is every indication that during the immediate future there will be an increasing number of applications of this type.

For the refiner, the introduction of mathematical methods promises to increase the profitability of his operations. For the economist, this area of research has a somewhat different interest. It gives him a more intimate working knowledge of the ancient problem of resource allocation, and provides him with a more satisfactory basis for estimating the production capabilities of a sector or indeed of an entire economy.

The current volume is intended to give the economist a sampling of the difficulties that must be faced when he attempts to apply his traditional categories to the problems of the industrial firm. It is seldom that he will come across a ready-made formulation of the production possibilities within a given plant — let alone an entire sector. On the contrary, he is likely to require the utmost ingenuity in order to construct a reliable quantitative picture with the data that are available to him.

As progress is made in the art of scheduling and the art of computing, it should be possible to include within the formal analyses an increasing number of the complexities of actual economic problems. At the same time, it is doubtful whether the intelligent observer will ever be fully satisfied with a current formulation. Reality will always appear to be oversimplified.

The outstanding advantage of a mathematical approach over an intuitive one is, not primarily that the idealized system can take account of more variables, but rather that the logical structure forces the analyst to bring out into the open those assumptions that otherwise lie hidden. Particularly in a large decentralized organization, each division tends to operate by using rules of thumb, and then by modifying the results in the light of experience. The more formal setup requires these assumptions to be made explicit. In addition, it furnishes a useful device for improving coördination between the various departments of the firm.

The introduction of mathematical economics is not a signal for the disappearance of business judgment or of engineering intuition within the refining industry. A formal study in this area inevitably depends upon individuals who have had experience with the manufacturing operations. There is no degree of mathematical subtlety that can rescue a faulty set of economic engineering assumptions.

2. A SUMMARY

This is a brief outline of the chapters that follow: first, a review of conventional refinery cost accounting methods; then, a problem involving the allocation of individual crude oils to several refineries at different locations; next, a typical intra-refinery conversion process — naphtha reforming; another intra-refinery activity — gasoline blending; and last, scheduling for a small integrated plant that performs cracking, recycling, and blending. For this full-scale refinery problem, not only a running plan is considered, but also a case involving the construction of new capacity.

As a result of the individual calculations, it is possible to obtain a better understanding of how a refiner might respond to changes in several kinds of variables: the market prices of refined products, the quality specifications of these products, and the amount of capital available for investment in new equipment. As far as the technology of the petroleum industry is concerned, no insurmountable difficulties are apparent. With patience and with some additional laboratory work, it should be possible to spell out the manufacturing process activities in whatever detail is required. For the prediction of market behavior within a decentralized economy, the same optimistic outlook is not warranted. No matter how much engineering realism is injected, the economist is still a long way from understanding the hierarchy of goals within a large private corporation. To a significant extent, the market behavior of such an institution depends upon the relative importance attached to such conflicting aims as the maximization of short-run profits, the preservation of good customer relations, and a long-term continuance of growth. Until a model of market behavior allows quantitatively for some of these factors, it is quite likely to prove sterile.

Petroleum executives do not consistently act as if they could market unlimited additional quantities of their products at the prevailing realizations, nor as if they could raise additional amounts of new capital at the prevailing rate of return. Paradoxically enough, the economic studies of long-term investment prepared within the major oil companies are typically based upon just such assumptions as these. The most satisfactory explanation of this tendency seems to be that the current market prices and rates of return on capital are used for estimating ideal shifts in policy, but that refiners do not attempt to move over instantaneously to these ideal positions. As in a servomechanism, there are continual discrepancies between the actual position and the one that currently looks optimal. In moving over to the new position, account must also be taken of the reaction of rival concerns and of the company's balance sheet position.

This process of adjustment is no longer within the realm of the refinery engineer, but rather that of the top management. Perhaps the most revealing comments along these lines are to be found in a letter written to the author several years ago by Dr. Robert E. Wilson, chairman of the board of the Standard Oil Company (Indiana):

What we do as a practical matter is to figure not only average costs, but incremental costs which represent essentially the cost of producing the last increment of any given product under the conditions that a particular refinery happens to be running. This incremental cost may be either above or below the average, depending on the cost of bringing in the last increment of crude and the type of equipment available for handling it.

While these incremental costs are a general guide as to what products we should be making and selling, there are always other obligations to be taken into account, and particularly obligations, in times of short supply, to keep all our customers supplied with their needs, even though we make somewhat less on one product than another. It is, therefore, not possible to follow rigidly the conclusions to which incremental cost data would point, but they are a guide to our refining and sales programs.

Dr. Wilson's reply will please neither the logical purist who maintains that business enterprises *always* follow an incremental cost rule, nor the confirmed institutionalist who insists that they *never* do. Either of these extreme positions is a gross oversimplification of actual practice. In a large corporation, just as in a governmental body, it is almost impossible to trace the decision-making process through all the individuals who participate in it. The fact remains that petroleum refiners do compute "out-of-pocket" costs, and at least look at these estimates when they map out their sales strategy. The studies contained within this volume do not pretend to answer the truly difficult questions of marketing and of finance, but rather to furnish part of the information that is needed in making these choices. It still remains for the executive group to perform the major policy decisions.

3. ORIENTATION OF THE BOOK

In order to keep down the amount of expository material contained within these pages, it is assumed that the reader is acquainted with (a) the solution of a set of simultaneous linear equations, and (b) the calculus problem of maximization under constraints.

Since the book centers upon the economic implications of mathematical models rather than upon the mechanics of computation, and since there are good descriptions generally available, I have not dwelt extensively upon the simplex method of linear programming. Instead, Chapter III contains a procedure for working out numerical solutions to a special class of linear programming problems, and Chapter VI includes some observations upon the general simplex method.

In going about their daily business, unfortunately, the econometrician and the engineer each use not only basic English but also a language peculiar to his own trade. Worse yet, there are words like "cost," "competition," and "profits" that are used by both groups, but which take on distinctly different meanings in the two contexts. In order for this work to be intelligible to both audiences, it has been necessary to simplify certain of the technical details, and to go over others at great length. The specialist is urged to skim over these passages. For the benefit of the nonspecialist, a glossary of technical terms is presented in the Appendix immediately following.

Appendix to Chapter I. A Glossary of Technical Terms[1]

Algorithm: Any formal computing procedure.

Alkylate: Product obtained in the alkylation process. Chemically, it is a complex molecule of the paraffinic series, formed by the introduction of an alkyl radical into an organic compound.

Alkylation: A synthetic process for the manufacture of components for aviation gasoline.

Antiknock Agents: Chemical compounds which, when added in small amounts to the fuel charge of an internal-combustion engine, have the property of suppressing or at least of strongly depressing knocking. The principal antiknock agent which has been developed for use in fuels is tetraethyl lead. Iron carbonyl and aniline (and other aromatic amines) have had limited use.

API Gravity: Arbitrary scale for measuring the density of oils, adopted

[1] Most of the oil refining definitions here are quoted directly from the glossary in *Fundamentals of Petroleum*, U. S. Bureau of Naval Personnel, NAVPERS 10883 (1953), pp. 161–172.

by the American Petroleum Institute. Water is 10° API; gasoline approximately 55°–60°.

ASTM: The initials of the American Society for Testing Materials.

ASTM Distillation: A distillation test made on such products as gasoline and kerosene to determine the initial and final boiling points and the boiling range.

Barrel: Petroleum industry uses 42-gallon barrel as the standard barrel.

Bottoms: In a distilling operation, that portion of the charge remaining in the still or flask at the end of the run; in pipe stilling or distillation, the portion which does not vaporize.

Btu: Abbreviation for British Thermal Unit, a unit of heat commonly used in heat engineering. It is the amount of heat necessary to raise the temperature of one pound of water one degree Fahrenheit.

Catalyst: A substance which effects, provokes, or accelerates reactions without itself being altered.

Catalytic Cracking: A method of cracking in which a catalyst is employed to bring about the desired chemical reaction.

Centistoke: A unit for the measurement of viscosity.

Cetane Number: Diesel fuel ignitability performance measured by the delay of combustion after injection of the fuel. It represents a comparison of a fuel with standards which are cetane in alpha-methyl-naphthalene.

Coking: The process of distilling a charge of oil to coke. In the last part of a coking run on a shell still, the bottom of the still is at a red heat and most of the volatile matter is driven out, leaving the coke hard and dry.

Cracked Gas Oil: The gas oil formed as one of the products of a cracking reaction. It should not be confused with the term "gas oil cracking stock," one of the possible inputs *into* a cracking still; "cracked gas oil" is sometimes known as "catalytic gas oil" if the cracking process has involved the use of catalysts.

Cracking: High temperature treatment of a given material (usually termed the "cracking stock" or "charging stock"). In this process, the long molecules of the cracking stock are broken up, with the attendant formation of gasoline. Other reaction products are gas oils, residual oils, and various gases.

Distillation: Distillation generally refers to vaporization processes in which the vapor evolved is recovered, usually by condensation, and a separation effected between those fractions which vaporize and those which remain in the bottoms. (See fractional distillation.)

Elasticity of Supply (Demand): A measure of the response of supply (or demand) to a variation in the market price under static competitive conditions. This is defined as the percentage change in the amount supplied (or demanded) per unit percentage change in the price.

End Point (EP): The highest temperature indicated on the thermometer inserted in the flask during a standard laboratory distillation test. This is generally the temperature at which no more vapor can be driven over into the condensing apparatus.

Fractional Distillation (See Distillation): Fractional distillation implies the use of equipment for effecting a more complete separation between the low and high boiling components in a mixture being distilled than does the general term distillation. It is usually accomplished by the use of a bubble tower or its equivalent.

Gas Oil: Term originally used to mean oil suitable for the manufacture of illuminating gas. Now employed to designate an overhead distillate product with a boiling range intermediate between that of kerosene and residual fuel oil. The material is used as fuel for home furnaces and diesel engines and as a cracking stock. Also known as "distillate oil" or "middle distillate."

Gasoline: A volatile liquid hydrocarbon fuel generally made from petroleum.

Identity Matrix: A square array containing the number "1" in each position along the principal diagonal and zero in the off-diagonal positions.

Initial Boiling Point (IBP): The temperature at which the first drop of distillate falls from the condenser into the receiver in a standard laboratory distillation procedure.

Kerosene: A petroleum overhead fraction with a boiling range intermediate between that of gasoline and gas oil. Used as an illuminant, stove oil, and tractor fuel.

Kinematic Viscosity: The ratio of viscosity of a liquid to its specific gravity at the temperature at which the viscosity is measured.

Lagrangean Multiplier. A term occurring in the mathematical process of maximization under constraints. At a point indicating a solution to the constrained-maximum problem, the Lagrangean multiplier represents the increment in payoff obtainable by *violating* one of the constraint conditions by one unit.

Matrix: A rectangular array of numbers. Certain matrix operations may be performed that are analogous to those involving ordinary scalar numbers.

Middle Distillates: A generic term for kerosenes and gas oils.

Naphthas: Oils of low boiling range (80°F to 440°F), usually of good color and odor when finished. Sometimes refers to gasoline components and sometimes to special products, solvents, etc.

Octane Number: Term used to indicate numerically the relative antiknock value of automotive gasolines, and of aviation gasolines having a rating below 100. It is based on a comparison with the reference fuels, iso-octane (100 octane number) and normal heptane (0 octane number). The octane number of an unknown fuel is the volume percent of iso-octane with normal heptane which matches the unknown fuel in knocking tendencies under a specified set of conditions. Either the Motor Method or the Research Method may be used in determining octane rating of automotive gasolines; either the Aviation Method or Supercharge Method may be used in determining the octane rating of aviation gasolines. The test method employed *must* be reported with the octane rating.

Opportunity Cost: A method of imputing value to a resource that is being employed for a specific purpose. It represents the value of that item if transferred to the next most profitable alternative use.

Parameter: A coefficient that is held constant while performing one part of an analysis, but that is considered to be a variable for purposes of the analysis as a whole.

Payoff: The term that is being maximized in a particular analysis.

Polymerization: A process for uniting light olefins to form hydrocarbons of higher molecular weight.

Recycling: The re-use of cracked distillate products as a charge stock in the same cracking process.

Reduced Crude: The bottoms remaining from a distillation of crude oil.

Reforming: A process for converting low octane number naphthas or gasolines into high octane number products.

Reid Vapor Pressure: The measure of pressure exerted on the interior of a special container (Reid Vapor Pressure apparatus), under specified test conditions.

Residual Fuel Oils: Fuel oils which include either reduced crudes or viscous cracked residuum. Used as fuel for industrial heat and power and also for marine and locomotive boilers.

Residuum: The dark colored, highly viscous oil remaining from crude oil, after the more volatile portion of the charge has been distilled off.

Scalar: A quantity subject to the usual laws of arithmetic or algebra. This term is generally employed in order to distinguish such a quantity from a "vector."

Shadow Price: A special case of a Lagrangean multiplier. This is the incremental value of any individual commodity in an economic optimization problem.

Straight-Run Gasoline (Raw Gasoline): A gasoline which is obtained directly from crude by fractional distillation.

Tetraethyl Lead: A volatile lead compound, $Pb(C_2H_5)_4$, which, when added in small proportions to gasoline, increases the octane rating.

Thermal Cracking: The process of cracking by heat or by heat and pressure.

Vector: A *set* of scalar quantities. Certain vector operations may be performed that are analogous to those involving ordinary scalar numbers.

Visbreaking: A mild cracking process employed in order to reduce the viscosity of residual stocks.

Viscosity: The property of resisting flow or deformation.

Chapter II

Conventional Methods of
Refinery Economic Analysis

1. STANDARD COST ACCOUNTING FORMULAE

From the standpoint of cost analysis, the outstanding feature of a petroleum refinery is the prevalence of joint costs. For example, in the primary atmospheric distillation of a crude oil, numerous "straight-run" fractions are obtained in fixed proportions.[1]

These straight-run cuts may be blended into salable products, or may in turn be charged into various types of conversion units. Of these conversion operations, the most common is known as "cracking." By means of cracking processes, the refiner can transform the middle distillates — kerosenes and gas oils — into gasoline, refinery gas, and residual fuel oil. The process breaks up the larger and heavier molecules into smaller ones in order to produce compounds of higher economic value.[2] Here again, the cost ac-

---

[1] These fractions are usually classified by their boiling range. The most volatile fraction is termed "refinery gas"; the next is gasoline or naphtha, which boils between, say, 150° and 400°F; the next is kerosene, between 400° and 520°F; and then comes distillate oil (also known as "gas oil") between 510° and 720°F. The residue (the fraction boiling above 720°F) is termed residual fuel oil. See below, Table VI.1.

[2] An additional feature of the conversion process is the fact that the gasoline produced is of higher quality than straight-run gasoline. Cracked gasoline has a higher octane rating and is generally used to upgrade the straight-run product. The

countant is faced with the problem of devising a suitable method for allocating the joint costs of the charge stock and the cracking unit among the individual products of the cracking reaction.

Two of the better-known formulae available for estimating costs in a joint output process like petroleum refining are the "sales value allocation" method and the "by-product" method. Neither of these give fully satisfactory answers to the economist's (or to the businessman's) product-mix problems. McKee, a leading petroleum cost accounting authority, advocates the sales value method:

The most ideally correct method is based on the theory that the same rate of gross profit is earned on each product. In other words, the allocable costs are spread to the products on the basis of their relative selling value . . . It will be seen that this method does assume the same rate of gross profit on each product.[3]

To illustrate this procedure, he presents the following calculation of costs over a one-month period.[4]

Crude oil used — 100 barrels	$126.00
Refining expense	21.00
Total expenses	$147.00

	Production (gallons)	Average market price ($/gal)	Market value (dollars)	Per-cent	Allocated cost (dollars)	
					Total	Per gallon
Gasoline stock	1,050	0.13	136.50	67	98.49	0.094
Fuel oil	2,730	.02	54.60	27	39.69	.015
Kerosene	84	.03	2.52	1	1.47	.017
Other products	252	.04	10.08	5	7.35	.029
Refining loss	84	—	—	—	—	—
Total	4,200	—	203.70	100	147.00	—

octane rating of cracked material is further enhanced by employing the newer catalytic methods rather than the older thermal processes.
[3] R. W. McKee, *Handbook of Petroleum Accounting* (New York, 1938), p. 316.
[4] McKee, p. 335.

Admittedly this method has the virtue of simplicity of calculation, but it does lack the advantage of being able to serve as a guide to management for price and output decisions. The calculations can be performed only after both of these decisions have been taken. It would be an obvious case of circular reasoning to suppose that cost estimates of this type can be used in order to help form these policies. They can only serve to indicate whether the operation as a whole is covering its expenses.

The by-product method is no more helpful. This system is based upon the curious premise that since gasoline sales constitute the most important source of revenue, the entire profit is made on this one product. Revenues from the other products only cover joint costs and really do not constitute profit. The "cost" of the by-products is their sales value at the prevailing market price. The total sales values of these products are subtracted from total refinery cost, and the remainder is the "cost" of producing gasoline. This formula may be illustrated by the following example, using some of the data in the preceding calculation.

Total expenses, as before		$147.00
Less market value of products		
other than gasoline		
Fuel oil	$54.60	
Kerosene	2.52	
Other products	10.08	
Total credits		67.20
Cost of 1,050 gallons of gasoline		79.80
Cost per gallon of gasoline		0.076

This gasoline cost figure of 7.6¢ per gallon may be compared with that of 9.4¢ computed via the sales realization method. It will be noted immediately how the by-product system imparts a downward bias to the "cost" of gasoline. Indeed, were profits sufficiently large, the by-product method could imply a *negative* cost for the main product. This system also constitutes an *ex post* method, and can only provide the refiner with a "break-even point" for the main product.

Operators seem to be well aware of the limitations of cost accounting for these purposes. An O.P.A. industry-wide survey on cost accounting methods concludes: "The refiners have apparently not based their selling prices on cost (i.e., cost as determined by cost accounting procedures) to any great extent, but rather have had to let competitive conditions and demand for their product determine their prices." [5]

During the era of the Temporary National Economic Committee a preliminary questionnaire on product costs was sent out to twenty major oil companies. Of the eighteen that replied at all, eight indicated that they used the sales realization method, and five that they employed the by-product formula.[6] All appeared skeptical about the figures that they did submit. The typical attitude is summed up by the reply of the Socony-Vacuum Oil Company.

We have no means of knowing, or even approximating except on a most arbitrary basis, what a gallon of gasoline "costs" our marketing department when it comes from our own refinery and is only one of many multiple products made from a single barrel of crude. It is true that, for income tax purposes and the ascertainment of surplus for dividend purposes, we must establish some "cost" for our products . . . For this inventory purpose, we use a method which is primarily the sales realization method . . .[7]

This, along with comments by other companies in a similar vein, would point to an important conclusion: cost figures on

[5] *A Report on Cost Accounting in Industry*, Accounting Department, United States Office of Price Administration (June 30, 1946), p. 41.

[6] *T.N.E.C. Hearings*, Part 17-A, pp. 10033–10113. One concern, the Texas Corporation, indicated that both the sales realization and the by-product methods were used within its organization.

[7] *Hearings*, Part 17-A, p. 10035.

individual products are not used primarily for purposes of internal control but rather for inventory valuation in financial and tax accounting. Under these circumstances, there is little to choose between the by-product and the sales realization method. Price fluctuations aside, both state the total value of inventories at the total cost of those inventories. The economist certainly cannot dispute this use with the accountant — provided that the accountant bears in mind the original purpose for which the cost figures were drawn up.

2. GASOLINE REPLACEMENT VALUE ESTIMATES

If not these standard costing techniques, what methods really are employed by the oil refiners for control purposes? To an economist, the procedure reported to the T.N.E.C. by two of the largest companies — the Standard Oil Company (New Jersey) and the Standard Oil Company (Indiana) — would appear the most familiar. The "gasoline value" formula amounts to what is frequently termed an "opportunity cost" calculation. It consists of (1) determining the value of gasoline by a by-product method, assuming that the refining operations are aimed at maximum conversion to gasoline; then (2) calculating the cost of kerosene and distillate oil as the value in terms of their potential gasoline and residual oil yields. In his testimony before the T.N.E.C., Dr. R. E. Wilson outlined the procedure in some detail.[8] For purposes of illustrating the method, however, it seems more convenient to use the accompanying figures condensed from a 1947 internal memorandum of one of the major companies.

(1) Processing crude for maximum gas-
    oline
Yields:
    Gasoline   64.0 percent
    Residual oil   37.6
Assumed prices per 42-gallon barrel:
    Crude oil $2.070
    Residual oil   1.680

[8] *Hearings*, Part 15, pp. 8646–8647.

Cost of gasoline plus fuel oil per barrel of crude:

Delivered cost of crude	$2.070		
Estimated cost of processing crude for maximum gasoline yield	0.612		
Total		$2.682	
Less price received for by-product fuel oil (37.6 percent × 1.680)		0.631	
Cost of 0.640 barrels of gasoline			$2.051
Cost per gallon of gasoline			0.0763

(2) Cracking heating oil for maximum gasoline

Yields:

Gasoline	65.5 percent
Residual oil	37.5

Costs per gallon of heating oil:

Value of gasoline obtainable on cracking heating oil stock (65.5 percent × 0.0763)	$0.0499		
Value of residual oil obtainable on cracking heating oil stock (37.5 percent × 1.68 ÷ 42)	0.0150		
Total value of products:		$0.0649	
Less cost of cracking heating oil stock and finishing, storing, and shipping products		0.0164	
Cracking value of heating oil stock			$0.0485
Plus cost of finishing heating oil stock into finished heating oil			0.0019
Cost of finished heating oil			0.0504

To re-state the procedure in the economist's language, middle distillates are valued at the opportunity cost of converting them into gasoline and residual oil, appropriate corrections being made for incremental refinery expenses. The primary objection to this

procedure is the assumption that cracking is necessarily conducted for maximum gasoline yield. If, instead, the reaction is
operated for less than the maximum overall conversion — e.g., if
the charging capacity is a bottleneck — the gasoline replacement
method will not lead to a good estimate of incremental value for
the cracking stock.

Despite its shortcomings, this costing principle appears to be
the most useful current technique for assisting decision-making
at the top management level. Refiners are well aware that it is
arbitrary to assume just one processing method as the alternative
to the sale of a particular oil, but in the absence of a better
technique, they consider that a replacement value estimate is of
considerable use in planning over-all sales policies.

3. COMPARISON OF MULTIPLE ALTERNATIVES

The primary drawback of the formula just described is the
assumption that there is only a single alternative to blending the
particular oil into a salable product. In fact, for making production decisions at the level of an individual plant, it is customary to compare a whole group of such alternatives. In Table II.1,
for example, there is reproduced a summary of the economics of
six possible schemes for a new refinery processing 10,000 barrels
per calendar day (B/CD) of Texas crude oil. The analysis was
performed by a group within the Houdry Process Corporation,[9]
and seems representative of the industry's practice.

Six basic setups are considered, and flow diagrams for the first
two are reproduced in Figure 1:

1. *Thermal* This processing scheme consists of conventional two-
coil thermally cracking the 40% vol. East Texas residuum to an ultimate yield of motor gasoline.
2. *Vacuum-distillation* The 40% vol. East Texas residuum is
vacuum-distilled to produce an 8.0% vol. asphalt. The vacuum gas oil
plus the excess light virgin gas oil are catalytically cracked and the
catalytic cycle stock remaining after blending the tar bottoms to No. 6
fuel oil specification is thermally cracked.

[9] G. F. Hornaday, N. D. Noll, C. C. Peavy, and W. Weinrich, "Various Refinery
Applications of Houdriflow Catalytic Cracking," presented at Western Petroleum
Refiners Association annual meeting, San Antonio, Texas, March 28–30, 1949,
Petroleum Refiner (June 1949).

Processing Method	$/B	1 Thermal B/CD	1 $/CD	2 Vacuum distillation, once through cat. cracking B/CD	2 $/CD	3 Visbreaking, once through cat. cracking B/CD	3 $/CD	4 Coking, once through cat. cracking B/CD	4 $/CD	3A Visbreaking recycle cat. cracking B/CD	3A $/CD	4A Coking, recycle cat. cracking B/CD	4A $/CD
REVENUE													
10 lb. Reid Vapor Pressure motor gasoline	4.830	6,126	29,589	6,694	32,332	6,797	32,830	7,370	35,597	6,745	32,578	7,120	34,390
Diesel fuel	3.570	1,000	3,570	1,000	3,570	1,000	3,570	1,000	3,570	1,000	3,570	1,000	3,570
No. 2 fuel	3.570							826	1,239	505	1,803	629	2,246
No. 6 fuel	1.50	2,378	3,567	1,781	2,672	1,623	2,435	858	1,287	1,024	1,536	243	365
Dry gas (fuel oil equivalent value)	1.50	739	1,109	744	1,116	794	1,191	80	84	753	1,130	794	1,191
Excess butane (as fuel)	1.05			43	45	65	68			124	130	130	136
Coke	5.00$/T								220				220
Total revenue			37,835		39,735		40,094		41,997		40,747		42,118
OPERATING COSTS													
Crude	2.95	10,000	29,500	10,000	29,500	10,000	29,500	10,000	29,500	10,000	29,500	10,000	29,500
Crude distillation		10,000	772	10,000	772	10,000	772	10,000	772	10,000	772	10,000	772
Houdriflow cracking				4,157	703	4,272	812	4,767	815	8,544	1,063	9,534	1,093
Gas plant			238		592		605		654		629		679
Thermal cracking		4,960	1,089	1,552	489	1,614	500	2,083	585				
Thermal reforming		1,990	453	1,990	453	1,990	453		453	1,990	453	1,990	453
Vacuum distillation				4,000	355								
Coking								1,497	427			1,497	427
Catalytic polymerization		261	226	358	256	373	261	422	276	373	261	407	272
Tetraethyl lead required for 80 octane number (F-2 test) (liters)	0.22¢/ml	437	962	210	464	228	502	337	742	147	324	203	447
Gasoline inhibitor		6,241	6	6,694	7	6,797	7	7,370	7	6,745	7	7,120	7
Taxes, interest, and insurance			401		647		589		695		586		678
Houdriflow royalty	0.05			4,157	208	4,272	214	4,767	238	4,272	214	4,767	238
Thermal royalty	0.03	6,950	209	3,542	106	3,604	108	4,073	122	1,990	60	1,990	60
Catalytic polymerization royalty	0.21	261	55	358	75	373	78	422	89	373	78	407	85
Total operating costs			$33,911		$34,627		$34,401		$35,375		$33,947		$34,711
Earnings ($/CD)			3,924		5,108		5,693		6,622		6,800		7,407
Payout time (years)			2.04		2.53		2.07		2.10		1.72		1.83
INVESTMENTS (PROCESS EQUIPMENT ONLY)													
Crude distillation			$ 700,000		$ 700,000		$ 700,000		$ 700,000		$ 700,000		$ 700,000
Houdriflow catalytic and feed preparation					1,505,000		1,550,000		1,600,000		1,900,000		1,960,000
Vacuum unit					500,000								
Gas plant			680,000		775,000		800,000		840,000		825,000		850,000
Thermal cracking and reforming			1,180,000		800,000		800,000		880,000		400,000		400,000
Catalytic polymerization and feed treating			370,000		440,000		450,000		485,000		450,000		470,000
Coker									570,000				570,000
Total			$2,930,000		$4,720,000		$4,300,000		$5,075,000		$4,275,000		$4,950,000

• Source: G. F. Hornaday, N. D. Noll, C. C. Peavy, and W. Weinrich, "Various Refinery Applications of Houdriflow Catalytic Cracking," *Petroleum Refiner* (June 1949), Table I.

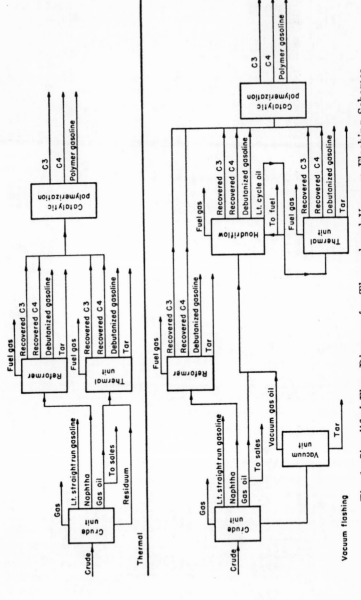

Fig. 1. Simplified Flow Diagram for Thermal and Vacuum Flashing Schemes

3. *Visbreaking* The 40% vol. East Texas residuum is charged to a flash tower and the resulting bottoms is visbroken and returned to the common flash tower. The flash-tower overhead stock plus excess virgin gas oil are catalytically cracked and the catalytic cycle stock remaining after blending the tar bottoms to No. 6 fuel oil specification is thermally cracked.

4. *Coking* The 40% vol. East Texas residuum is flash-distilled to obtain an overhead fraction and the resulting bottoms is coked. The combined flash tower overhead and total coker distillate plus excess light virgin gas oil are catalytically cracked and the catalytic cycle stock is thermally cracked.

. . . two additional processing schemes were calculated to show the effect of recyle catalytic cracking instead of thermally cracking the cycle stock. The two additional situations are alternates for the visbreaking and coking processing schemes and are designated 3A and 4A.[10]

As a matter of strict logic, it does not seem appropriate to consider these six possibilities alone, and to exclude all others. For example, why assume that there is just one catalytic reactor temperature worth considering? Why assume that thermal cracking is the only alternative to the recycling of gas oil produced in the catalytic cracker? Why assume that it is always preferable to sell 1,000 B/CD of gas oil directly as diesel fuel rather than employ it as a cracking stock? When needled on such issues, an engineer will usually reply as follows: "You can't hope to make a formal analysis of all conceivable processing schemes. On the basis of your past experience, you concentrate on the individual ones that seem most promising. It is expensive to obtain any kind of pilot plant information for these economic calculations, but you cannot make reliable forecasts of product yields without such numbers. These six cases alone require many man-hours to be spent in computing."

Apparently two distinct issues are involved here. One is the basic limitation of the refining data that are available. The other is the awkwardness of performing calculations that require many interlocking variables. It has been the object of the present research to develop actual examples where the economist can suggest improved computing schemes, despite the limitations of the

---

[10] Hornaday, *et al*, pp. 2–3 of reprint.

data that are on hand. Four instances of such applications will now be given. In each of these, it was possible to utilize information that is now being collected on a routine basis by oil companies. In two of the four studies — the gasoline blending problem and the recycling problem — it was necessary to employ an automatic electric calculator. In the other two instances, it was possible to employ hand computing methods, and still effect an improvement over the techniques actually being used.

Not only should an economist be in a position to suggest more systematic ways of organizing the data, but his training ought also lead him to question the performance criteria that are frequently employed within the industry. In Table II.1, for example, process 3A, visbreaking and recycle catalytic cracking, leads to the lowest payout time — 1.72 years. Process 4A, coking and recycle catalytic cracking, leads to the highest absolute level of earnings — $7,407 per calendar day — and process 1, thermal cracking, requires the lowest initial outlay. It is not at all obvious that process 3A, with the least payout time, is the best bet here. If a refiner has limited financial resources, he may prefer alternative 1, and if he has $4,950,000, the 4A scheme may still provide the best investment opening for him.

A comptroller might well be interested in determining the difference in estimated earnings between a $3,500,000 initial investment and a $4,500,000 one, but it is quite difficult to rearrange the information in Table II.1 for such purposes. In view of the initial assumption of a fixed crude oil input of 10,000 B/CD, it is not legitimate to suppose that the only alteration is in the scale of process 3A, and that the payout time will remain the same. The fact that the refiner decides to use a visbreaker and a thermal cracker for the cycle stock need not commit him to an investment of exactly $4,275,000. Specific investment limitations can easily be included within a linear programming analysis, and an example of this type is given below in Chapter VI.

Chapter III

A Crude Oil Allocation Problem[1]

1. INTRODUCTION

One of the most complex scheduling problems in the U. S. petroleum industry arises in connection with the assignment of crude oils between the various refineries of a multiplant firm. For example, the Esso Standard Oil Company's East Coast Group alone includes five plants — Linden and Bayonne, New Jersey; Baltimore, Maryland; Everett, Massachusetts; and Charleston, South Carolina. During any one period of time, these units together are likely to be processing between one hundred and two hundred different types of crude oils. Some of these come from Texas, others from Louisiana; still others will come from Venezuela and Colombia. Not only do the crude oils vary in physical and chemical composition, but also they differ markedly from one another in the field costs of production and in the costs of transportation to any one of the potential refining centers. The plants themselves have a wide variety of processing equipment. One, for example, may have a catalytic cracking unit,

[1] The subjects discussed in this chapter have many points of contact with those in a paper by A. Charnes, W. W. Cooper, and B. Mellon, "A Model for Programming and Sensitivity Analysis in an Integrated Oil Company," *Econometrica*, vol. 22, no. 2 (April 1954), pp. 193–217. Rather than dwell upon the similarity of approach, three points of divergence will be mentioned. (1) The scheme proposed here would tie in closely with a current method of crude oil evaluation. (2) Several distinct refinery locations are considered. (3) The Hitchcock-Koopmans-Dantzig computing algorithm is employed rather than Charnes's duality theorem approach.

and for this reason be able to secure a significantly higher yield of premium grade gasoline than a plant that contains only thermal refining equipment.

Under these circumstances, anyone would be rash to claim that he could determine *the* ideal way to schedule crude shipments to individual refineries. In order to come even close to this objective, a substantial amount of new empirical data would be required. Then, too, if the problem were handled as a general linear programming system, the computational requirements would be large. Were the analyst required to take the full complexities of geography into account, it would be difficult to avoid building up a matrix that contained less than 500 rows. If, on the other hand, one is willing to make certain kinds of simplifying assumptions, it turns out that the mathematical form of the crude oil allocation problem fits directly into that of the Hitchcock-Koopmans "empty vessel" example.[2] There are several pragmatic advantages in setting up the oil scheduling problem according to this form. The data requirements can be supplied out of numbers which companies now collect, and in addition, the computing requirements are simple. Even for a case with ten refineries and two hundred crude oils, the problem could be solved within less than eight man-days — and without resorting to the use of automatic computing devices.

In reducing the problem to the Hitchcock-Koopmans form, one does not get something for nothing. In particular, the analysis does *not* take adequate account of intra-refinery bottlenecks, of finished product blending, and of market sales limitations; but on none of these counts does the model proposed here seem any less realistic than the present methods of crude oil evaluation. The basic data would be virtually the same. Of course, if the mathematical results were followed blindly in the one case, and some judgment were applied in the other, the purely mechanical approach would come off second best. There is no particular reason, however, for supposing that one method could not be supplemented by the other. Once an optimal schedule had been determined by the suggested algorithm, the results could always

---

[2] See T. C. Koopmans and S. Reiter, "A Model of Transportation," *Activity Analysis of Production and Allocation*, Cowles Commission Monograph 13 (New York, 1951), chapter XIV, pp. 222–259.

be modified through "seat-of-pants" estimates. By proceeding in this way, some account could be taken of the elements that are not formally handled within the computing framework — the equipment bottlenecks, the blending problems, and so on.

2. EXISTING METHODS OF CRUDE OIL EVALUATION

An official in one major company has described the costing procedures that are currently used in the following terms:

The relative crude evaluation consists of determining the value of one crude in relation to another. Such evaluations are made by technical groups for a number of crudes that may be available for processing in refineries. The procedure consists of processing on paper the crude in question through the sequence of refining steps representative of a given refinery, and blending the various streams to final saleable products. These products are credited at the going refinery realizations. The value of crude is determined by subtracting from these credits the cost of all manufacturing operations involved, including storing and shipping expenses. This value of the crude, when compared to its delivered price, indicates profit or loss to be sustained from processing it through this particular refinery. A comparison of this profit or loss with the corresponding figure for another crude will give the desired relative value.

The above crude evaluations can be carried out on an average basis when using average yields and total manufacturing costs, or on an incremental basis when using incremental product yields, qualities and out-of-pocket costs. Incremental yields, qualities, and costs represent the changes when crude throughputs are increased or decreased by a relatively small volume.

In order to make clear just what is involved in this method, a hypothetical calculation has been carried through for an individual crude-oil-refinery assignment. The problem is to consider the assignment of the ith crude oil to the jth refinery, and to trace through the various cost elements involved in taking this crude from the producing field, transporting it to the refinery, processing it there, and then disposing of the finished products at the refinery gate. In order to distinguish these individual elements of the problem the following algebraic symbols are defined:

c_i,   cost ($2.00) of purchasing one barrel of the ith crude oil, and bringing it to the field pipeline

l_{ij},   cost ($0.75) of transporting one barrel of the ith crude oil to the jth refinery

m_{ij},   general manufacturing expenses ($0.50) associated with distilling one barrel of the ith crude oil at the jth refinery   (III.1)

y_{ijk},   volumetric yield (percent) of the kth product when operating on the ith crude at the jth refinery

r_{ijk},   realization (dollars) at the jth refinery's gate on one barrel of the kth product from the ith crude

e_{ijk},   manufacturing expense (dollars) allocated to one barrel of the kth product of the jth refinery when operating on the ith crude.

To consider a simplified case, suppose that there are only four products from the ith crude at the jth plant, and that the individual yields, realizations, and expenses are as given in Table III.1.

Table III.1. Hypothetical Refined Product Yields, Realizations, and Expenses on the ith Crude Oil at the jth Refinery

Index k	Product	Product yields, percent by volume of crude distilled, y_{ijk}	Refinery gate product realizations, r_{ijk} (dollars per barrel)	Manufacturing expenses allocated to individual products, e_{ijk} (dollars per barrel)
1.	Gasoline	40	6.00	0.30
2.	Kerosene	15	4.50	.10
3.	Heating oil	20	4.00	.05
4.	Residual fuel oil	24	2.50	.03
	Total liquid yield on crude[a]	99		

[a] The volume loss in refining is considered to be 1 percent here.

As the company official's memorandum suggests, the various cost factors, c_i, t_{ij}, m_{ij}, y_{ijk}, r_{ijk}, and e_{ijk}, must all be taken into account when considering the net company-wide effect of increasing or decreasing the shipment of one barrel of the hypothetical ith crude to the jth refinery. That is, the company's net realization per barrel on the i-to-j transaction, termed p_{ij}, may be defined as follows:

$$
\begin{aligned}
p_{ij} &= \sum_k y_{ijk}(r_{ijk} - e_{ijk}) - c_i - t_{ij} - m_{ij} \\
&= 0.40(6.00 - 0.30) + 0.15(4.50 - 0.10) + \\
&\quad 0.20(4.00 - 0.05) + 0.24(2.50 - 0.03) - \\
&\quad 2.00 - 0.75 - 0.50 \\
&= \$0.2322. \tag{III.2}
\end{aligned}
$$

This net unit value is determined by crediting the yield of each product with its refinery gate realization, debiting each product for the manufacturing expenses associated with it, and also debiting the costs of producing, transporting, and distilling one barrel of the crude.[3]

Up to this point, several features of the problem have been deliberately avoided. Of these, the most significant are the words "unit cost." Do these represent average costs or incremental ones? The last paragraph of the preceding memorandum makes it clear that both of these concepts are employed, and that each serves a legitimate purpose. If one is looking into the distant future and contemplating the purchase of additional facilities of a similar type, then surely it is appropriate to inquire whether the average costs can be covered over the service life of the equipment. If, on the other hand, the unit has already been purchased, the depreciation and interest charges are constant, whether or not the plant is operated. In the short run, therefore, it will pay to operate such a unit even if no more than the incremental expenses are covered. This problem comes down to the economist's familiar distinction between short-run and long-run marginal costs. Essentially, the petroleum company concerns itself with the historically observed average expenses because, in

---

[3] To facilitate the presentation here, it is assumed that there are no crude oil losses enroute from the field to the refinery.

default of other estimates, these provide at least some rough indications of what the long-run incremental costs might be.

In the calculation of p_{ij}, there naturally arises the question of joint *manufacturing* costs.[4] As indicated in the preceding chapter, cracking unit costs are involved whenever a heavy oil is converted into gasoline and other reactor products. Under these circumstances, it is quite arbitrary how the costs of operating the cracking unit are "split up" between the various end items. Fortunately, it turns out that this feature of the process does not raise insoluble difficulties. Suppose, for example, that in the preceding estimate of p_{ij}, the cracking unit costs had amounted to $0.30 per barrel of fresh feed. Suppose also that the processing scheme for the ith crude oil required that 60 per cent by volume of the crude be employed as cracking stock. It would then be necessary to deduct an additional amount from the net company realization, i.e., $0.60(\$0.30) = \$0.18$ cracking unit costs per barrel of crude distilled. Alternatively, the cracking material could be considered as a fictitious refinery product with yield $y_{ijk} = 0.60$, but bringing a zero r_{ijk} refinery gate realization, and entailing a manufacturing expense of $e_{ijk} = \$0.30$ per barrel of fresh feed. This individual joint cost problem does not create serious obstacles for present methods of crude oil evaluation.

3. A MATHEMATICAL ANALYSIS OF THE SCHEDULING PROBLEM

In setting up a formal model of the crude oil scheduling problem, the p_{ij} coefficients may be taken over directly from the conventional estimates of relative crude oil values. It is further assumed that there are certain upper limits, Q_i, on the total daily quantities available of each crude oil, and that there are certain distillation capacity limitations, D_j, associated with each refinery. The problem is that of determining the shipments of the various crude oils to each refinery in such a way as to maximize the over-all company realization. If the variable x_{ij} is employed to indicate the amount of the ith crude oil shipped to the jth refinery, and if there are m refineries and n crude oils, the problem reduces to the following constrained-maximum form:

[4] The joint costs connected with the crude oil *raw material* have already been brought into the analysis through the last three terms of equation (III.2).

Maximize $\qquad \sum_{ij} p_{ij}\, x_{ij},$

subject to: $\qquad \sum_{j} x_{ij} \le Q_i \qquad (i = 1, 2, \ldots, n),$

$$\qquad\qquad \sum_{i} x_{ij} \le D_j \qquad (j = 1, 2, \ldots, m),$$

$\qquad\qquad x_{ij} \ge 0 \qquad\quad \text{(all } i \text{ and } j\text{)}.$

(III.3)

In other words, the thing that is to be maximized is the *sum* of the net refinery realizations over all of the individual shipments. In grinding out this optimal schedule, there are three requirements imposed. First, for each of the n crude oils, the total shipments programmed must not exceed the crude oil availability limits, Q_i. Second, for each of the m refineries, the total crude runs planned must not go over the basic distillation capacities, D_j. And third — which will be obvious to any refiner — the schedule must not call for negative amounts of crude oil to be shipped.

To the uninitiated, the problem stated in (III.3) inevitably looks formidable. If, say, there were ten refineries in the system and two hundred individual crude oils, there would be 2,000 degrees of freedom! Luckily, the model fits into the "empty vessel" form, and there is a straightforward computing procedure available to handle this class of problems. Before going on to examine this algorithm, it is important to set out in explicit form the assumptions that underlie the particular formulation.

Of these simplifications, perhaps the most serious is the notion that the value of a given crude oil to a refinery is independent of the remainder of the "crude slate" available to that plant or to any of the others. It is assumed that for the market conditions prevailing at an individual refinery, there is a single standard way to process a given crude. There is a serious question of whether the intermediate streams produced from this crude are competing with other oils for the use of bottleneck facilities. There are similar interaction problems connected with blending intermediate streams from one crude oil with those from another. The analysis does not cover the question of shipping unfinished oils between refineries. It does not adequately handle the marketing problems that are faced by a major refiner. Although it is assumed that the company can dispose of any additional refined products at the prevailing realizations, this will seldom be fully realistic.

After enumerating all the qualifications above, it is natural to raise the question of whether the (III.3) model deserves any consideration at all. The best that can be said in its favor is that all of its defects coincide with those of the existing crude oil evaluation methods. And to the extent that the analysis includes explicit upper bounds on the company's crude oil availabilities and on its atmospheric distillation capacity, the procedure represents an improvement over those that are currently used. If, in the existing scheduling system, a refiner exercises his informed judgment and does not follow the "relative crude values" rigidly, this same expertise could also be applied to the tentative solutions turned out by a mathematical model. The computation is cheap; it imposes no extraordinary data requirements, and it provides some of the inputs that would be needed in handling larger and more realistic systems. By going over to a general linear programming approach in future models, it would be feasible to take account, not only of geography, but also of the intra-refinery bottleneck equipment, of the blending possibilities, of inter-refinery shipments, and of market limitations on finished products. Several of these component topics are treated in subsequent chapters, but the present volume does not pretend to handle the integrated company's full scheduling problem, following crude oil from the wellhead to refined products in the consumer's tank. Before attempting the more ambitious job, it seems worthwhile to inquire whether any modest improvements can be effected.

4. COMPUTATIONAL PROCEDURE[5]

4.1 OUTLINE OF THE PROCEDURE

Although, in principle, an example could be worked out for a situation covering two hundred crude oils and ten refineries, the

---

[5] In writing this section, I have benefited greatly from a manuscript by Robert Dorfman, "The Transportation Problem" (August 19, 1953). Dorfman's material will ultimately appear as one chapter in a book on linear programming written jointly by him, Paul Samuelson and Robert Solow.

system considered here will include only five crudes and three refineries. Once this simple fictitious case has been worked out, the underlying ideas can be readily extended to large-scale problems. The basic ingredients of the analysis are the crude oil availabilities, Q_i; the distillation capacities, D_j; and the net refinery realizations, p_{ij}. These coefficients are assumed to take on the values indicated in Tables III.2 and III.3. Since it is assumed that the company incurs no *incremental* costs in leaving crude oil idle, there are five zero net realizations listed for the individual crudes in Table III.3 under the column heading "surplus crude oil."

Under the conditions indicated by Tables III.2 and III.3, it is possible to make some preliminary statements on the properties

Table III.2. Hypothetical Company-wide Crude Oil Availabilities and Distillation Capacities

Q_1 = Quantity available of crude oil 1   =   39,000 B/CD
Q_2 = Quantity available of crude oil 2   =   32,000
Q_3 = Quantity available of crude oil 3   =   24,000
Q_4 = Quantity available of crude oil 4   =   15,000
Q_5 = Quantity available of crude oil 5   =   11,000
$\Sigma_i Q_i$ = Total crude oil available   = 121,000

D_1 = Atmospheric distillation capacity of refinery 1 =   50,000 B/CD
D_2 = Atmospheric distillation capacity of refinery 2 =   35,000
D_3 = Atmospheric distillation capacity of refinery 3 =   25,000
$\Sigma_j D_j$ = Total distillation capacity   = 110,000

Table III.3. Hypothetical Net Realization upon Crude Oil i at Refinery j, p_{ij}
(dollars per barrel)

	Refinery 1	Refinery 2	Refinery 3	Surplus crude oil
Crude 1	0.353	0.232	0.201	0
Crude 2	.407	.350	.276	0
Crude 3	.513	.475	.490	0
Crude 4	.326	.265	.249	0
Crude 5	.190	.117	.073	0

of an optimal routing schedule. The total amount of crude oil available — 121,000 (B/CD) — exceeds the total of distillation capacity — 110,000 B/CD. At least 11,000 B/CD of the crude, therefore, cannot be utilized in the refineries. At the same time, no matter how the optimal routings turn out, it will never pay to leave any of the distillation capacity idle. So long as there is crude available, and all of the net realizations are positive, there is always something to be gained by putting crude oil through an otherwise idle plant.

Although these special features were deliberately built into the illustrative problem there is no assurance that things will turn out this way in an actual operation. There may be some negative p_{ij} net realization coefficients, and there may be an excess of distillation capacity over the flow of crude available. True, either of these conditions might render a stockholder unhappy, but neither of them should cause any special difficulties to the programmer. As for the negative refinery realizations, he knows that he need never make these shipments, that he can always do better by leaving both the crude oil and the distillation capacity idle.[6] And if the scheduler faces an excess of total distillation capacity over the crude available, he can treat this problem symmetrically with that of an excess of crude oil. In the one case, he sets up a balancing item termed "surplus crude oil," an account to which any excess crudes can be allocated with zero profit or loss. In the other, he sets up an account called "surplus distillation capacity," and again a zero net gain is associated with all items in this category.

If one agrees to the convention of giving "surplus crude oil" the subscript "4," the routing problem now becomes a matter of selecting x_{ij} entries in Table III.4. By definition of the balancing

---

[6] "Never" is an excessively strong word to employ. If, for example, there is a *lower* limit on the amount of a given crude that must be used, some of that oil will have to be put through the refineries, regardless of negative net company-wide realizations. The lower limit might arise out of the fact that a certain field would flood if pumping ceased. The lower limit might also come out of a contractual commitment to take the material, or out of the desire to maintain good relations with a local regulatory body. In any of these situations, the scheduling problem can be handled by imposing a special *upper* bound upon the amount of idle crude oil or of idle distillation capacity.

items, the row entries must sum up exactly to the total amount of each crude available. And since it has been previously determined that it will be unprofitable to leave any distillation capacity idle, the five entries in each column must sum up to the capacity totals, D_j.

Table III.4. Schedule of x_{ij}, Crude Oil Allocations (B/CD)

	Refinery 1	Refinery 2	Refinery 3	Surplus crude oil	Total amount available of crude i
Crude oil 1	x_{11}	x_{12}	x_{13}	x_{14}	$Q_1 = 39,001$
Crude oil 2	x_{21}	x_{22}	x_{23}	x_{24}	$Q_2 = 32,001$
Crude oil 3	x_{31}	x_{32}	x_{33}	x_{34}	$Q_3 = 24,001$
Crude oil 4	x_{41}	x_{42}	x_{43}	x_{44}	$Q_4 = 15,001$
Crude oil 5	x_{51}	x_{52}	x_{53}	x_{54}	$Q_5 = 11,001$
Total distillation capacity, refinery j	$D_1 =$ 50,000	$D_2 =$ 35,000	$D_3 =$ 25,000	$D_4 =$ 11,005	$\Sigma_j D_j = \Sigma_i Q_i =$ 121,005

The conditions imposed by Table III.3 correspond precisely to those of the restraints of equations (III.3). As before, none of the entries may be negative ones. The x_{ij} "activities" in the columns must sum up to the distillation capacities for the three plants, and the individual rows must account for the total allocation of each of the five crudes. The problem remains one of selecting the x_{ij} in such a way as to maximize the over-all company realization, $\sum_{ij} p_{ij} x_{ij}$.

The careful reader will notice that each of the five Q_i given in the table exceeds the exact amount of crude available by one barrel. Similarly, the item D_4 is listed as five barrels in excess of the exact amount of surplus crude oil indicated previously. These small overages cannot distort the final answer, for the optimal schedules will always be rounded off to the nearest significant digit — in this instance, to the nearest 1,000 barrels. The one-barrel "epsilon factors" are merely an aid in the process of computing, but they do not influence the end results of the calculation. They are introduced to avoid any problems of "degeneracy." [7] Rather than go into details here on this particular difficulty, it will suffice to state without proof the following theorem (due to A. Orden): let d equal the least significant digit in the totals D_j and Q_i. If there are m refineries and n crude oils, and if $n \geq m$, then the "epsilon factor" may be any number smaller than $d/2n$.[8] If, for example, there were two hundred crude oils under consideration, and the totals were stated to the nearest ten barrels, any number could be taken as the "epsilon factor" so long as it remained smaller than $\dfrac{10}{2 \times 200} = 0.025$ barrels. The coefficient ϵ must be added to each of the Q_i totals, and $n \cdot \epsilon$ to one of the D_j. In this way, it is possible to prevent any *partial* sum of the D_j from being equal to a partial sum of the Q_i.

Through the artifice of the epsilon factors, one can guarantee the three following properties of the operating schedule. (1) There will be *at least* $(m + n - 1)$ positive x_{ij} in any schedule that satisfies the row- and column-total restrictions.[9] Such a program is termed a "feasible" one. If a feasible program contains *exactly* $(m + n - 1)$ positive x_{ij} activities, then it is called a "basic feasible" schedule. The set of $(m + n - 1)$ activities actually employed is termed the "basis." (2) The optimal x_{ij} schedule will always be a "basic feasible" solution. In case there are several alternative schedules that are optimal, i.e., that maximize $\sum_{ij} p_{ij} x_{ij}$, at least one of these will be a basic solution. (3) For

[7] The reader who is interested in this special topic is urged to consult George B. Dantzig, "Application of the Simplex Method to a Transportation Problem," *Activity Analysis of Production and Allocation*, chapter XXIII, pp. 365–367. Another treatment is to be found in A. Charnes, "Optimality and Degeneracy in Linear Programming," *Econometrica*, vol. 20, no. 2 (April 1952), pp. 160–170.

[8] Dantzig, p. 366.

[9] Why should there be no more than $(m + n - 1)$ positive levels of individual shipments required? Since there are m refineries, including the "surplus crude oil" account, there are m distillation capacity restrictions (column totals) that must be satisfied. In addition, there are n crude oils, and the total allocation of each of these materials must be accounted for. (The crude oil allocations are controlled through the individual row totals.) Now since *both* the row and the column totals are specified in advance, there is one less independent condition than $(m + n)$, and the system will contain $(m + n - 1)$ simultaneous linear equations. If the conditions of the problem can be satisfied at all, an optimum solution will not require more than the stated number of independent variables.

each basic feasible set of x_{ij}, there may be defined a group of $(m + n - 1)$ independent dual variables u_i and v_j. For each x_{ij} activity *in* the basis, the dual variables are chosen in such a way that $p_{ij} = u_i + v_j$. Once an optimal basis is reached, these dual variables will exhibit a useful property. At such an optimum, for each x_{ij} activity *outside* the basis, $p_{ij} \leq u_i + v_j$.

In the crude oil allocation example under study here, we know that there is no need for more than $(m + n - 1) = (4 + 5 - 1)$ $= 8$ activities operated at positive levels in the optimal routing plan. All of the remaining 12 possibilities can be set equal to zero.

Associated with the eight x_{ij} in any basic solution will be the eight dual variables u_i and v_j. One can think of u_i as the incremental payoff associated with the ith crude oil that is available. For the given basis of x_{ij}, this is a "shadow price" for the ith crude — a value that would leave the refiner exactly indifferent whether to take on or to give up an incremental barrel of this material.

By symmetry, v_j can be considered as the value imputed to an incremental barrel per day of distillation capacity in the jth refinery. For each x_{ij} activity within the basis, this value is set at such a level that the refiner would find the direct payoff associated with p_{ij} exactly offset by the "imputed indirect costs" of the particular crude oil and distillation capacity $(u_i + v_j)$.

For the x_{ij} in an optimal basis, the direct payoff will be exactly cancelled out by the "shadow costs." In addition, for those x_{ij} activities that lie *outside* the optimum program, the direct p_{ij} payoff will never exceed $(u_i + v_j)$. The reason? Whenever, for a given basis, there is a non-included activity that yields a "surplus," the programmer will always be able to increase $\sum_{ij} p_{ij} x_{ij}$. The existence of a positive $(p_{ij} - u_i - v_j)$ is a signal to him that the given basis is not an optimal one; for if the activity is introduced into the existing schedule by the amount θ_{ij}, he will find that the total payoff increases by an amount $\theta_{ij}(p_{ij} - u_i - v_j)$. The greater the level at which θ_{ij} can be set, the larger will be the improvement in the company profits.

There is, of course, a limit on this process of generating payoff by means of increasing θ_{ij}. If the one crude oil allocation is made sufficiently large, there will be some other activity previously in the basis that inevitably becomes squeezed down to a zero level.[10] In other words, a new basis has been generated by introducing a profitable activity, deleting a less profitable one, and juggling the other shipments around in such a way as to preserve the row- and column-totals. This new basis is again tested for optimality, and the process of improving the net company payoff is continued until an optimum schedule is finally determined. Since there are only a finite number of bases, and since the payoff is being improved at each step, the process of computation must finally come to an end. It does not go on indefinitely, and will generally terminate after approximately $(m + n - 1)$ base shifts.[11]

4.2 THE NUMERICAL EXAMPLE

Recalling the theory of the preceding section, we note that there are three distinct steps involved in solving for the optimum allocation schedule. The first is that of obtaining a basic feasible solution. The second is that of determining the dual variables associated with the particular basis. And the third is that of employing the dual variables in such a way as to indicate whether the given basis is an optimal one, or, if not, which of the previously excluded activities should be brought into the basis. The computation process alternates between the second and the third steps until finally an optimum set of activities has been reached.

Although in principle the selection of the initial basis is a purely arbitrary choice, it will generally be possible to reduce the ultimate number of base shifts by starting out from a relatively favorable position. One good way to start the process is to operate under the principle of "absolute advantage." That is, referring back to Table III.3, the most profitable-looking activity appears to be x_{31}. For every barrel of Crude 3 shipped to Refinery 1, the company nets $0.513.

It *looks* as though the maximum profit will be realized by making allocation x_{31} as large as possible. What is the upper limit on this shipment? Is it the distillation capacity of Refinery 1, or is it the availability of type 3 crude oil? Going to

---

[10] The term "degeneracy" is applied to a case in which two or more of the former activities are *simultaneously* reduced to the zero level by the introduction of the new one. The epsilon technique serves to rule out situations of this type.

[11] Dantzig, p. 373.

Table III.4, it is easy to see that the ceiling is imposed by the crude oil availability, for $Q_3 = 24,001$, while D_1 would permit a level of 50,000 B/CD. As soon as these things are determined, it is easy to see that the three potential alternative uses of Crude 3 must be set at zero. These three are now automatically excluded from the first basis. Table III.5 indicates the manner in which Table III.4 is modified to take account of the assignment of 24,001 B/CD to x_{31}. To indicate that the activity is included within the basis, an asterisk is placed in the upper right-hand corner of box 31.

Table III.5. Schedule of x_{ij}. First Phase in Determining a Basic Feasible Solution (B/CD)

	Refinery 1	Refinery 2	Refinery 3	Surplus crude oil	Total remaining amount of crude i available
Crude oil 1	x_{11}	x_{12}	x_{13}	x_{14}	39,001
Crude oil 2	x_{21}	x_{22}	x_{23}	x_{24}	32,001
Crude oil 3	24,001*	0	0	0	(24,001 − 24,001) = 0
Crude oil 4	x_{41}	x_{42}	x_{43}	x_{44}	15,001
Crude oil 5	x_{51}	x_{52}	x_{53}	x_{54}	11,001
Total remaining distillation capacity, refinery j	(50,000 − 24,001) = 25,999	35,000	25,000	11,005	

* The asterisk indicates that the activity is included within the basis.

After bringing the most profitable-looking activity into the basis, what is to be done next? Table III.3 indicates that number 33 would be the best choice, for its p_{ij} exceeds that of all the others except p_{31}. Unfortunately, this second-best step cannot be taken. All of Crude 3 has already been used in Refinery 1, and activity 33 is automatically excluded from the starting basis. Activity 32 seems next in order of profitability, but it is excluded for the same reasons.

Following 32 in order of absolute advantage is the activity 21.

The shipment of a barrel of Crude oil 2 to Refinery 1 is associated with a payoff of \$0.407. This activity has not previously been excluded, and it is now introduced to the maximum extent possible. The programmer will find that the upper limit on this allocation is set by the 25,999 distillation capacity remaining in Refinery 1, and not by the 32,001 B/CD availability of Crude 2. Taking account of these facts, Table III.5 becomes transformed into III.6.[13] Since the introduction of activity 21 fills up the entire capacity of the first refinery, the potential activities 11, 41, and 51 are automatically dropped from the initial basis.

Continuing in this spirit, the refiner will find it possible to bring six additional activities into the basis. In descending order of absolute profitability, these are x_{22}, x_{42}, x_{12}, x_{13}, x_{14}, and x_{54}. Once the total of eight activities has been chosen, the first step is completed — the selection of a *feasible* basis. (This basis, although feasible, is not necessarily an optimum one.) The eight individual crude oil assignments are all listed in Table III.7.

Following the choice of a feasible solution, the programmer moves on to determine the eight dual variables, u_i and v_j. The first phase here consists of noting down those eight p_{ij} that are associated with the eight activities operating at positive levels. This information is displayed in Table III.8.

The eight dual variables, i.e., shadow prices, are defined in such a way that the refiner's direct realization on each of the eight p_{ij} is exactly offset by the "indirect costs," u_i and v_j. Although Table III.8 lists nine dual variables altogether, evidently one of these constitutes a superfluous degree of freedom. Any individual out of this group may be set at an arbitrary value. Since column 2 contains three entries (i.e., three crudes are shipped into the second refinery), it is convenient to assign a zero level to v_2, the shadow price of distillation capacity in Refinery 2.

Once this arbitrary selection is made, the values of each of the other shadow prices can be read off in a chain-like sequence. In general terms, the scheduler must solve the following system of eight simultaneous linear equations:

$$p_{ij} = u_i + v_j \quad (ij \equiv 12, 13, 14, 21, 22, 31, 42, 54) \qquad \text{(III.4)}$$

[13] For the purpose of making the various shifts, a blackboard or a transparent template turns out to be a valuable accessory.

Table III.6. Schedule of x_{ij}. Second Phase in Determining a Basic Feasible Solution (B/CD)

	Refinery 1	Refinery 2	Refinery 3	Surplus crude oil	Total remaining amount available of crude i
Crude oil 1	0	x_{12}	x_{13}	x_{14}	39,001
Crude oil 2	25,999*	x_{22}	x_{23}	x_{24}	32,001 — 25,999 = 6,002
Crude oil 3	24,001*	0	0	0	24,001 — 24,001 = 0
Crude oil 4	0	x_{42}	x_{43}	x_{44}	15,001
Crude oil 5	0	x_{52}	x_{53}	x_{54}	11,001
Total remaining amount available of capacity, refinery j	(50,000 — 24,001 — 25,999) = 0	35,000	25,000	11,005	

* The asterisk indicates that the activity is included within the basis.

Table III.7. Schedule of x_{ij}. Initial Basic Feasible Solution (B/CD)

	Refinery 1	Refinery 2	Refinery 3	Surplus crude oil	Total amount available of crude i
Crude oil 1	0	13,997*	25,000*	4*	39,001
Crude oil 2	25,999*	6,002*	0	0	32,001
Crude oil 3	24,001*	0	0	0	24,001
Crude oil 4	0	15,001*	0	0	15,001
Crude oil 5	0	0	0	11,001*	11,001
Total distillation capacity, refinery j	50,000	35,000	25,000	11,005	

* The asterisk indicates that the activity is included within the basis.

Table III.8. Net Company Realizations, p_{ij}, and Dual Variables, u_i and v_j. Initial Basic Feasible Solution

	Refinery 1	Refinery 2	Refinery 3	Surplus crude oil	"Shadow price" of crude oil i, u_i ($/barrel)
Crude oil 1		0.232*	0.201*	0*	u_1
Crude oil 2	0.407*	.350*			u_2
Crude oil 3	.513*				u_3
Crude oil 4		.265*			u_4
Crude oil 5				0*	u_5
"Shadow price" of distillation capacity, refinery j, v_j ($/barrel)	v_1	v_2	v_3	v_4	

* The asterisk indicates that the activity is included within the basis.

The feature that makes the Hitchcock-Koopmans-Dantzig method an easy one is that it will *always* be possible to rearrange the equations in such a way that they fall into a triangular form. In actual computation practice, the system may be read off directly from such a table as III.8, but for illustrative purposes here, the full explicit sequence will be solved. In the present instance, one can ascertain the values of u_1, u_2, and u_4 directly, and then use these to determine the other unknowns. The system moves along as follows:

$$
\begin{array}{lll}
1) & v_2 = 0 & \text{(arbitrary)} \\
2) & u_1 = p_{12} - v_2 = \$0.232 \text{ per barrel} \\
3) & u_2 = p_{22} - v_2 = & .350 \\
4) & u_4 = p_{42} - v_2 = & .265 \\
5) & v_1 = p_{21} - u_2 = & .057 \\
6) & u_3 = p_{31} - v_1 = & .456 \\
7) & v_3 = p_{13} - u_1 = & -.031 \\
8) & v_4 = p_{14} - u_1 = & -.232 \\
9) & u_5 = p_{54} - v_4 = & .232
\end{array}
\qquad \text{(III.5)}
$$

Now that the dual variables have been determined, they can be employed for testing the optimality of the basis. For each of the twenty possible x_{ij} activities, the programmer forms the expression $(p_{ij} - u_i - v_j)$. If this expression is negative, it indicates that the indirect costs of introducing the activity ij would outweigh any direct gains. If, on the other hand, this form is positive, the refiner knows that he can improve the overall payoff by introducing the activity.

According to Table III.9, the eight basic activities display a zero level of surplus, as is required. The table also shows that

Table III.9. Dual Variables, u_i and v_j; Activity Surpluses $p_{ij} - u_i - v_j$. Initial Basic Feasible Solution (dollars per barrel)

	Refinery 1	Refinery 2	Refinery 3	Surplus crude oil	u_i
Crude oil 1	0.064	0*	0*	0*	0.232
Crude oil 2	0*	0*	−0.043	−0.118	.350
Crude oil 3	0*	0.019	.065	−.224	.456
Crude oil 4	.004	0*	.015	−.033	.265
Crude oil 5	−.099	−.115	−.128	0*	.232
v_j	.057	0	−.031	−.232	

* The asterisk indicates that the activity is included within the basis.

there are seven more that have negative surpluses, i.e., that these shipments would diminish the total company payoff. And finally, there are five activities that yield a positive surplus of direct gain. These are the following ones: 11, 32, 33, 41, and 43. If any of these are introduced, there will be an improvement in the overall return.

Which of the five is now to be brought in? If the activity x_{ij} yields a positive surplus, and may be introduced up to a maximum level of θ_{ij}, the improvement in over-all return would be:

$$\Delta \sum_{ij} p_{ij} x_{ij} = \theta_{ij}(p_{ij} - u_i - v_j) \qquad \text{(III.6)}$$

Since the programmer wants to make the largest immediate increase in payoff, ideally he should select the ijth activity that would maximize expression (III.6). Computationally, however, it is easier to bring in the activity associated with the maximum *surplus*, $(p_{ij} - u_i - v_j)$. This non-ideal procedure has the advantage of requiring a single computation for θ_{ij}, rather than several trials — one for each of the activities displaying a positive surplus.

In the case at hand, the maximum indirect gain is provided by activity 33, the shipment of Crude 3 to Refinery 3. For each barrel that is sent over this routing, the over-all company profits will improve by $0.065. The refiner sets $x_{33} = \theta_{33}$, and tries to make θ_{33} as large as possible. As the third step in the optimization procedure, he proceeds to modify Table III.7, p. 37, the previous routing schedule. He inserts the *unknown* variable θ_{33} in box 33 of Table III.10. To offset the positive row entry, he finds that he must reduce by θ_{33} the allocation of Crude oil 3 to Refinery 1. For this reason, activity 31 is set at the level $(24,001 - \theta_{33})$. Next, in order to keep Refinery 1 filled up with crude oil, the scheduler must increase the allocation of Crude 2 by θ_{33} barrels. This change now reduces the amount of Crude 2 that is available for Refinery 2. The shift is offset by bringing more of Crude 1 into the second refinery, and then by reducing the amount of Crude 1 that goes into the third refinery.

Table III.10. Modifications Required in Initial x_{ij} Shipping Schedule in Order to Introduce x_{33} (B/CD)

	Refinery 1	Refinery 2	Refinery 3	Surplus crude oil	Total amount available of crude i
Crude oil 1	0	$13,997 + \theta_{33}$*	$25,000 - \theta_{33}$*	4*	39,001
Crude oil 2	$25,999 + \theta_{33}$*	$6,002 - \theta_{33}$*	0	0	32,001
Crude oil 3	$24,001 - \theta_{33}$*	0	θ_{33}	0	24,001
Crude oil 4	0	15,001*	0	0	15,001
Crude oil 5	0	0	0	11,001*	11,001
Total distillation capacity, refinery j	50,000	35,000	25,000	11,005	

* The asterisk indicates that the activity is included within the basis.

The sequence is now complete, for the positive and negative values of θ_{33} cancel out, thereby preserving the row and column totals. In order to introduce activity 33 into the basis, there had

to be a six-way shift of crude oils. The only allocations that remained unaffected were 14, 42, and 54.

In order to calculate the benefits of the shift, the scheduler
must first determine how large θ_{33} can be set. He will note that
there are three boxes in Table III.10 where the unknown θ_{33} is
prefixed by a negative sign — activities 13, 22, and 31. Any of
these three could conceivably set an upper limit on this unknown.
In the present case, it turned out that activity 22 had been
smaller than either 13 or 31. It is apparent that when θ_{33} is set at
a level of 6,002 B/CD, the previous shipment of Crude 2 to
Refinery 2 is reduced to zero, and activity 22 may be deleted
from the basis. The new basis retains seven of the eight activities
from before, but it includes one new route pattern — activity 33
in place of number 22.

If we agree to neglect the epsilon factors, and round off the
activity levels to the nearest 1,000 barrels, we find that the six-
way switch has increased the company's payoff from $37,242 to
$37,632, an initial improvement of $390 per day. This change
may also be calculated as follows:

$$\theta_{33}(p_{33} - u_3 - v_3) = 6,000 \text{ B/CD } (\$0.065/\text{B}) = \$390/\text{CD}.$$
$$\text{(III.7)}$$

The third step in the computing sequence is now finished, and
the programmer can go back to the second — the determination
of values of the dual variables for the new basis. The procedure is
identical with what was done in Table III.8 and equations
(III.5). One of the u_i or v_j is arbitrarily set at zero, and the others
are determined in such a way that the activities in the *second*
basis are associated with a zero surplus. Having calculated the
shadow prices connected with the second basis, the refiner can
see which of the excluded activities will yield a positive surplus.
In the present example, activity 11 turns out to be the most
attractive. Knowing that activity 11 is to be introduced, the
refiner goes on to the third step — determining the maximum
allowable value for θ_{11} as in Table III.10. He finds that, in order
to make θ_{11} as large as possible, it is necessary to knock activity 31
out of the basis. Again, payoff is improved, this time by $2,322
per day.

So much for the principle of "absolute advantage." The shipment of Crude oil 3 to Refinery 1 initially *looked* like a good thing, but after a short computation no longer seems so attractive. Refinery 3 has a "comparative advantage" in the utilization of this material, and none of it is shipped to Refinery 1.

Even at the new level of payoff, the persevering scheduler does not rest. When he determines the shadow prices for the third basis, he again finds that he has a nonoptimal solution. By reintroducing activity 22 and by deleting number 12, he is able to effect another improvement.

Table III.11. Shipping Schedules, x_{ij}, for Seven Successive Base Shifts, Rounded off (10^3 B/CD)

				Basis Number			
ij	I	II	III	IV	V	VI	VII
11	0	0	18	38	39	39	39
12	14	20	20	0	0	0	0
13	25	19	1	1	0	0	0
14	ϵ	ϵ	ϵ	ϵ	ϵ	0	0
21	26	32	32	12	11	11	0
22	6	0	0	20	21	21	32
23	0	0	0	0	0	0	0
24	0	0	0	0	0	0	0
31	24	18	0	0	0	0	0
32	0	0	0	0	0	0	0
33	0	6	24	24	24	24	24
34	0	0	0	0	0	0	0
41	0	0	0	0	0	0	11
42	15	15	15	15	14	14	3
43	0	0	0	0	1	1	1
44	0	0	0	0	0	ϵ	ϵ
51	0	0	0	0	0	0	0
52	0	0	0	0	0	0	0
53	0	0	0	0	0	0	0
54	11	11	11	11	11	11	11
Payoff, $\sum\limits_{ij} p_{ij}x_{ij}$ (dollars per day)	37,242	37,632	39,954	41,234	41,313	41,313	41,357

He keeps alternating between the shadow prices of the second step and the θ_{ij} variable of the third step until he has gone from the initial basis to a seventh one. The rounded-off x_{ij} for each of the iterations are reproduced in Table III.11. Finally, at the seventh basis, the programmer learns that he can rest upon his laurels. After he determines the dual variables for this stage, he sees that there are no activities that yield positive surpluses (Table III.12). Given the crude oil availabilities, the distillation capacities, and the net refinery realizations of the model, he can devise no program that will lead to a higher payoff for the company than $41,357 per day. He has come up from an initial level of $37,242, an improvement of $4,115 per day. In a year's operation of the three refineries, this discrepancy would amount to $1.5 millions, a non-negligible sum of money.

Table III.12. Dual Variables, u_i and v_j; Activity Surpluses $p_{ij} - u_i - v_j$. Optimum Solution, Seventh Basis (dollars per barrel)

	Refinery 1	Refinery 2	Refinery 3	Surplus crude oil	u_i
Crude oil 1	0*	−0.060	−0.075	−0.027	0.027
Crude oil 2	−0.004	0*	−.058	−.085	.085
Crude oil 3	−.054	−.031	0*	−.241	.241
Crude oil 4	0*	0*	0*	0*	0
Crude oil 5	−.136	−.148	−.176	0*	0
v_j	.326	.265	.249	0	

* The asterisk indicates that the activity is included within the basis.

5. SUMMARY AND CONCLUSIONS

This section has presented an admittedly simplified analysis of the allocation of crude oils between the various units of a multi-plant integrated concern. The model skips over many features that are of real significance to a refiner — the omission of all manufacturing bottlenecks other than primary distillation, the neglect of the interdependence that comes from blending together intermediate oils from several crude sources, and the assumption that additional amounts of the end products may be marketed at a constant realization. No account is taken of the interchange of unfinished oils between refineries, and no attempt is made to choose between alternative operating conditions on the various conversion units. In principle, these elements of the problem all hang together, and should not be treated in isolation from one another.

Having scored all these points against the model, the reader should recall the initial aim — to produce an analysis that is at least one step better than the formal ones currently employed. Not only does the mathematical technique take notice of the net realization estimates that are available in large integrated companies today, but it also handles the question of crude oil availabilities and of primary distillation capacities. The Hitchcock-Koopmans-Dantzig procedure ensures that the numerical analysis can be both rapid and inexpensive.

Chapter IV

A Naphtha Reforming Problem

1. INTRODUCTION

Although it is customary for refinery economics problems to be handled through the methods described in Chapter II (that is, by comparing several distinct alternative processing schemes), there are instances in which men in the industry have employed more formal techniques of analysis. As an example of such methods, it is instructive to study a report by Nelson R. Adams and George D. Creelman of the M. W. Kellogg Company, "A Comparison of the Economics of Single Pass Naphtha Reforming with the Use of Tetraethyl Lead" (April 30, 1935).[1] Despite the fact that this analysis covers refinery equipment that is now obsolescent, and despite the large changes that have occurred in the price and cost relationships, the basic report remains of substantial general interest. The *methods* of economic analysis employed in the 1935 naphtha reforming study are still representative of the Kellogg Company's practice, and for this reason alone deserve attention. The report illustrates the type of thinking that

[1] The full report is available only in mimeographed form. The basic results, however, have appeared in an article by W. W. Gary and N. R. Adams, "Economics of Reforming and Leading Mid-Continent Gasoline," *National Petroleum News* (Refinery Technology edition, August 11, 1937), pp. R-83–90.

is required when planning for any major departure from an existing refinery layout.

The processing scheme considered here is summarized in the flow diagram, Figure 2. The problem is one of selecting the operating conditions and the scale of the reforming equipment in such a way as to maximize the refinery realization that can be obtained from a given volume of a straight-run naphtha. The optimization revolves around two basic independent variables, X and Y, where X represents the volume of heavy naphtha

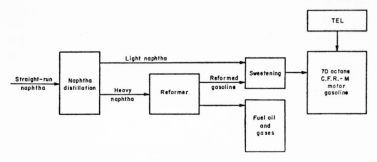

Fig. 2. A Schematic Diagram for Thermal Reforming and Ethyl Fluid Blending

charged to the reformer, and where Y represents the output of reformed gasoline as a proportion of the feed to the reformer. In addition to the gasoline, the process produces oil and dry gases that can be used as refinery fuel, and these are assigned the same value as the equivalent volume of heavy fuel oil. The total gasoline product is to be raised to an octane number of 70 C.F.R.M. by a combination of reforming and of adding tetraethyl lead (TEL). The more reformer gasoline yield the operator is willing to sacrifice, the greater will be the octane rating of the reformed product and, therefore, the lower will be the amount of ethyl fluid that is consumed. The refiner must consider not only the credits obtainable from the gasoline and fuel products, but also the operating costs associated with the reformer and with ethyl fluid and gasoline sweetening.

In presenting this material, the first step will be to indicate the underlying assumptions, the second to explain the graphical method of solution devised by Adams and Creelman, and the final one to suggest an alternative method of handling the problem algebraically.

The algebraic method seems to provide a less expensive computation scheme than the graphical one, but this is not in itself a commanding advantage. So long as the analysis is confined to two independent variables, as in the case here, the algebraic method does not really come in its own. But when it becomes necessary to consider a full-scale refinery problem with many interlocking decisions, algebraic methods provide the only tools for a formal analysis. In order to set the background for the refinery-wide problems considered in later chapters, the naphtha reformer unit is here studied in isolation. All of the technical and economic assumptions are those originally made by Adams and Creelman.

2. BASIC ECONOMIC ASSUMPTIONS

2.1 MARKET CONDITIONS

It is assumed that, per unit of time, there are available 100 gallons of 100°–400° boiling range Mid-Continent straight-run naphtha. The total amount of this material is fixed, but it may be processed by numerous alternative combinations of ethylizing and of thermal reforming. The refinery realization is termed C, and this is the quantity that is to be maximized. The variable, C, is the algebraic sum of (1) gasoline revenues, (2) fuel gas credits, (3) ethyl fluid costs, (4) reformer operating costs, and (5) gasoline sweetening costs.

Supposedly there are no limitations effective upon the total amount of gasoline or of fuel gas that may be produced. Regardless of the scale of output, these are credited at the prevailing refinery realizations. In addition, ethyl fluid is taken to be available at the going rates. Since this evaluation of market conditions is the one originally put into the 1935 analysis, there is no need to discuss it at length here. It is likely to be realistic for the smaller refiners, but not necessarily so for the major ones. (See below, Chapter V, part 5.1.) In the case of a larger company, it may be particularly important to include within the mathematical model an explicit limitation on the total amount of gasoline that may be marketed at the going prices.

The economic assumptions may be summarized as follows. Starting with 100 gallons of naphtha, the refiner is free to vary the amount of reforming and of ethyl fluid blending so long as the gasoline product meets a C.F.R.M. octane specification of 70. The more that is charged to the reformer, and the more severe the reforming treatment, the lower will be the TEL requirements. The gasoline and fuel products are to be credited at the going rates, regardless of the amounts produced. Not only the yields of gasoline and fuel but also the costs of ethyl fluid, of reforming, and of gasoline sweetening will be determined by the two independent variables, X and Y.

2.2 ECONOMIES OF SCALE IN REFORMING OPERATIONS

Since the analysis concerns the installation of a new reforming unit, one of the variables subject to control is the charging

Table IV.1. Single Pass Naphtha Reforming Costs

	Case A	Case B	Case C
Gravity of fresh feed	45° API	45° API	45° API
Barrels per stream day, fresh feed	1000	5000	10,000
Barrels per calendar day, fresh feed	900	4500	9,000
Time efficiency	90%	90%	90%
Operating cost per calendar day:			
Operating labor	$ 57.80	$ 90.20	$103.20
Operating supplies	5.40	27.00	54.00
Maintenance	30.60	62.30	95.40
Fuel at 60¢ per barrel	21.60	108.00	216.00
Steam at 24.5¢/10^3 lb	44.00	39.40	63.70
Power at 0.54¢/Kwh	2.70	20.20	35.50
Water at 0.66¢/10^3 gal	6.10	29.00	56.50
General reforming expense	13.30	22.90	29.80
Total per calendar day (excluding fixed charges)	181.50	399.00	654.10
Cost per barrel charge	0.2015	0.0887	0.0727
Cost per gallon charge	0.00480	0.00211	0.00173
Approximate total investment	279,000	568,000	870,000
Approximate investment per B/CD of fresh feed	310	126	97

capacity of that unit. This capacity must be sufficient to accommodate X gallons of heavy naphtha per unit of time. Apparently there are economies of scale in the construction and operation of reformers, and for this reason the authors felt it worthwhile to get some notion of the importance of the scale factor.

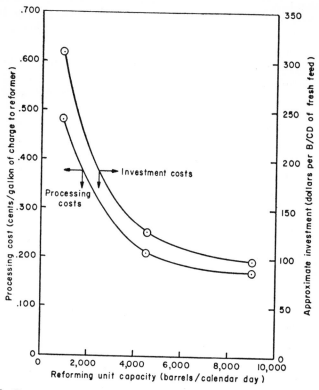

Fig. 3. Processing Costs and Initial Investment Costs vs Reformer Unit Size. Single Pass Reforming of Virgin Heavy Naphthas

Both the capital investment and the operating costs were calculated for three levels of input capacity: 900 B/CD; 4500 B/CD; and 9000 B/CD. The operating charges do not include royalties, interest or depreciation charges on the capital investment. The cost estimates (in 1935 dollars) are reproduced in Table IV.1, and are summarized in Figure 3.

From this figure, two results are apparent. In moving from a unit with a capacity of 900 B/CD to one of 4500 B/CD, there is a large decline in unit costs, both for the capital investment and for the operating charges. By contrast, in moving from a level of 4500 to one of 9000 B/CD, there is only a modest reduction in these unit costs. In fact, using these three capacity levels, the draftsman faired in curves which seem to approach asymptotes at unit cost levels that are comparable to those of a 9000 B/CD installation.

Evidently, for reformer sizes of 4500 B/CD or more, operating costs remain in the neighborhood of 0.2¢ per gallon of charge to the reformer. In order to simplify the body of the analysis, therefore, Adams and Creelman take the reformer costs at this constant level, but, in addition, they provide a correction factor to take care of situations where these depart significantly from the 0.2¢ per gallon rate. For the remainder of this discussion, the problem of scale will be ignored. It is assumed that, regardless of the choice of operating conditions, the *variation* in reforming costs would not be sufficiently large to have a feedback influence upon this choice.

Capital costs, as distinct from the current operating charges, were not incorporated directly within the analysis. Adams and Creelman reason as follows:

It is felt that optimum reforming conditions should always be calculated ... on this basis, with fixed charges and royalty costs to be figured only in the final stage of the cost comparison. The reason for this lies in the wide variation in methods of amortization of investment items covering equipment and paid up royalties ... Fixed charges and paid up royalties enter the picture only in determining whether or not a given refiner should reform at all and not in determining the optimum way to reform once the capital has been invested for reforming equipment and processing rights.[2]

Not all members of the refining industry or of the economics profession would agree to apply the doctrine of "bygones are bygones" to a prospective *new* plant. Nevertheless, the mathematical model has been set up so as to correspond to this principle.

[2] "Economics of Single Pass Naphtha Reforming" (1935), p. 28.

3. BASIC ENGINEERING ASSUMPTIONS

Before deriving the equations that govern the output of gasoline, fuel gas, etc., Adams and Creelman listed a number of simplifying assumptions that they had made. These were as follows:

First, evaporation and sweetening losses shall be disregarded.

Second, the octane and leading characteristics of a gasoline whether sweetened or unsweetened shall be considered identical.

Third, sweetening cost unless otherwise specified shall be assumed constant at 0.05¢/gallon of gasoline sweetened.

Fourth, virgin gasoline of 100–400°F. boiling range shall require no treatment other than sweetening and leading to produce 70 CFR-M octane finished gasoline.

Fifth, no limitation of allowable tetraethyl lead concentration in finished gasoline shall be considered as having any bearing on ethylizing vs. reforming economics.

Sixth, tetraethyl lead shall be considered to cost 0.30¢/cc and the cost of blending gasolines and mixing with ethyl fluid shall be neglected.

Seventh, no differential evaluation due to varying vapor pressure or boiling range characteristics of finished gasolines shall be made.

Eighth, approximate evaluation of charging stocks shall not take into account lead credit or debit due to effective overall changes in lead consumption resulting from blending with extraneous refinery gasolines. Approximate charging stock evaluations shall be based on the assumption that the leading requirements of a blend of gasoline can be estimated from a weighted average of the lead requirements of the individual components raised to the same octane number.

(Note: Approximate charging stock valuations are to be distinguished from the ultimate values which allow for corrections involving the factors of casinghead price, refinery butane balance and lead susceptibility of the total gasoline produced in the refinery.)[3]

Since Adams and Creelman have put down their postulates so clearly, it is not necessary to comment at great length upon these. From the standpoint of the general reader, it is the fifth, seventh and eighth which are of most interest. To rephrase these points, even though the naphtha operations are to be conducted within

[3] "Economics of Single Pass Naphtha Reforming," pp. 14–16.

an integrated refinery that handles other gasoline blending stocks, it is assumed that the *blending* problem does not influence the selection of optimal naphtha reforming conditions. Although the refinery's final blended gasoline must meet certain requirements as to the maximum vapor pressure, the maximum lead level, and the minimum volatility, these restrictions are left out of account in choosing the preferred quantity of reformer feed and the proportion of gasoline yield from that material.

If these assumptions are reasonable ones, it is possible to divorce the reforming and the blending problems from one another, and to handle them separately. (That is the way the matter has been approached in this chapter and the one immediately following.) If, on the other hand, this interdependence cannot be ignored, then it is necessary to construct a mathematical system that includes both phases of the refinery's operations. The linear programming model of a simplified thermal cracking refinery (Chapter VI below) represents one step in the direction of bringing the two analyses together. Unless there is a formal link established, it may be extremely difficult to distinguish between "approximate charging stock valuations" and "the ultimate values which allow for corrections involving the factors of casinghead price, refinery butane balance and lead susceptibility of the total gasoline produced in the refinery."[4] In any event, before proceeding to consider the interdependence between the reforming, cracking, and blending problems, it is essential to obtain insight into the workings of each of these smaller blocks. Without first examining the elementary operations in some detail, it is useless to hope that an aggregate model will give meaningful answers to production problems at the refinery-wide, company-wide, or industry-wide level.

4. PRODUCT YIELDS AND INPUT COSTS

In order to set up their relationships for graphical solution, Adams and Creelman make the following definitions:

[4] There is a close relation between the vapor pressure specification of a motor fuel and the amounts of casinghead gasoline and of butane which may be included in the finished blend. Both of these components, although of high octane rating, have vapor pressures which exceed the permissible level in a salable product.

X = Number of gallons of heavy naphtha charged to the reformer per unit of time;

Y = Reformer gasoline yield, proportion by volume of the fresh heavy naphtha feed;

C = Value of 100 gallons of *total* straight-run naphtha (cents);

C_V = Value of one gallon of raw virgin *light* naphtha (cents per gallon) of light naphtha;

C_R = Value of one gallon of raw virgin *heavy* naphtha (cents per gallon) of heavy naphtha;

L_V = Cost of TEL required to bring one gallon of light virgin naphtha to 70 octane rating (cents per gallon of light virgin naphtha);

L_R = Cost of TEL required to bring one gallon of reformed gasoline to 70 octane rating (cents per gallon of reformed gasoline);

G = Refinery realization on 70 octane, sweetened gasoline (cents per gallon);

F = Refinery fuel value (cents per gallon);

P = Reformer operating cost (cents per gallon of fresh feed to the reformer);

S = Sweetening cost (cents per gallon of sweetened gasoline).

Of the variables defined above, the first two are the independent ones X and Y. The five immediately following — C, C_V, C_R, L_V, and L_R — are all dependent upon these two. The remaining four are considered to be parameters, and not subject to control within the framework of the naphtha reforming problem. The gasoline and fuel credits, as well as the reformer and sweetening costs, are all taken to be constants.

From the flow diagram, Figure 2, it can be seen that the realization per gallon of light virgin naphtha, C_V, amounts to the revenues on a gallon of finished gasoline minus the unit costs of sweetening and of adding ethyl fluid. This is expressed in the following identity equation:

$$C_V = G - S - L_V. \qquad \text{(IV.1)}$$

Similarly, per gallon of reformer gasoline product, the realization amounts to $(G - S - L_R)$. For each gallon of reformer charge, the refiner receives Y gallons of gasoline yield and also $(1 - Y)$ gallons of fuel yield. In addition, he incurs an operating cost of P. Pricing out these three elements and combining them into an identity equation for the net value per gallon of reformer charge stock, we have

$$C_R = Y(G - S - L_R) + (1 - Y)F - P. \qquad \text{(IV.2)}$$

The quantity to be maximized is C, the total realization on the 100 gallons of straight-run naphtha. On each of the X gallons of heavy material, it has been determined that the refiner receives C_R. On each of the $(100 - X)$ gallons of light material, he receives C_V. Multiplying the unit realizations by the total quantities produced, and adding, it is found that

$$C = X C_R + (100 - X) C_V. \qquad \text{(IV.3)}$$

Substituting terms from (IV.1) and (IV.2), equation (IV.3) becomes:

$$C = X[Y(G - S - L_R) + (1 - Y)F - P] \\ + (100 - X)[G - S - L_V]. \qquad \text{(IV.4)}$$

The entire problem is now reduced to equation (IV.4). This relation includes the constant terms related to the gasoline and fuel realizations, the sweetening costs, and the operating costs. There are cross-product terms associated with the two degrees of freedom X and Y, and there are also present the two dependent variables, L_R and L_V, the ethyl concentration levels. An apparatus of seven interlocking nomograms is then set up to deal with the problem of maximizing equation (IV.4) simultaneously with respect to X and Y. The nomograms are reproduced in schematic form as Figures 4a through 4g.

5. THE GRAPHICAL SOLUTION

Basically, the graphical procedure provides for a trial-and-error solution. First an arbitrary value is assumed for X, the quantity of reformer charge. Next, taking X as given, the optimal value of Y may be determined. With this information in hand, the refiner can evaluate C_V and C_R, and then from equation (IV.3), he can compute the optimum total naphtha realization, C, for the *particular* value of X that was originally assumed. In turn, he proceeds to assume another value for X, and evaluates the payoff that corresponds to it. He continues to vary X by small increments until, by trial and error, he has selected that combination

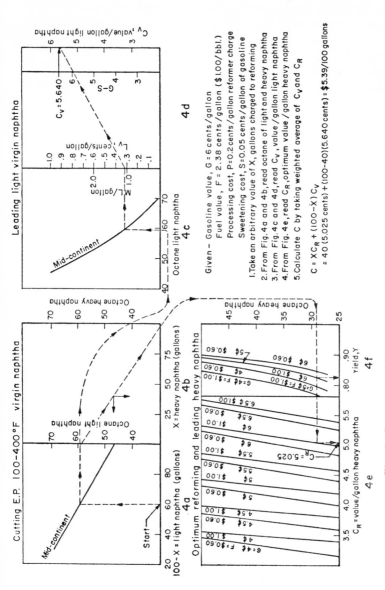

Fig. 4. Summary of Nomograms, Naphtha Reforming Problem

of X and Y which maximizes the over-all return on the 100 gallons of raw naphtha. The nomograms are arranged in such a way that they allow several alternative values to be attached to G and F, respectively, the realizations on gasoline and fuel. In all cases, however, it is assumed that the reformer processing cost, P, amounts to 0.2¢ per gallon of charge; that the sweetening costs

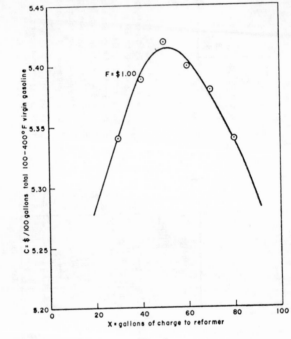

Fig. 4g.

are 0.05¢ per gallon of gasoline; and that ethyl fluid costs are 0.3¢ per milliliter.

The detailed steps required to use the nomograms are as follows:

(1) Take an arbitrary value for X, for example a 40-gallon charge of heavy naphtha to the reformer.

(2) From Figure 4a, read off the octane number for the light material that results from the 40–60 split. In this case, the octane rating of the light material comes to 59.6.

(3) From Figure 4c, determine the number of milliliters required to bring the light naphtha to 70 octane (that is, 1.03 ml per gallon).

(4) Entering the multiplication and subtraction nomogram, Figure 4d, with this ethyl fluid requirement, and with $(G - S) = 5.95¢$, read off the value of C_V, 5.64¢ per gallon. This part of the job corresponds to equation (IV.1).

(5) Returning now to Figure 4b, begin with the two points — one corresponding to an octane rating of 59.6 for the light naphtha, and the other corresponding to 47.3, the rating of the total initial material. Assuming that octane ratings blend linearly,[5] the nomogram indicates a knock index of 29 for the heavy naphtha.

(6) Enter Figures 4e and 4f with the octane rating of the heavy cut and determine the combination of G and F that represents the closest approximation to the current gasoline and fuel prices. With this information, the optimal yield of reformer gasoline, Y, may be determined from 4f, and the corresponding unit realization on heavy naphtha, C_R, from 4e. Assuming that gasoline is worth 6.0¢ and fuel 2.38¢ per gallon ($1.00 per barrel), Y turns out to be 0.83 gallons of gasoline yield per gallon of reformer charge. The unit value of heavy naphtha may be read off as 5.025¢ per gallon.

(7) Calculate C, the total payoff, by substituting in equation (IV.3) the term C_V determined in step (4), and C_R from step (6). This total comes to $5.39 for the 100 gallons of starting material, assuming that the reformer feed is set at 40 gallons per unit of time.

(8) On a separate chart, Figure 4g, now plot the value of $5.39 versus the abscissa of 40 gallons of reformer charge. Then repeat steps (1) through (7) for other values of X: 30, 50, 60, 70 and 80 gallons of charge. Plotting all these points on Figure 4g, it can be seen that the optimal amount of reformer charge will lie in the neighborhood of 50 gallons. The corresponding value of Y represents a reformer gasoline yield of 0.84, and the optimal realization, C, comes to $5.42 per 100 gallons of raw naphtha.

With the seven nomograms, Figures 4a through 4g, the refiner is enabled to solve equation (IV.4) for a wide variety of price

[5] See Chapter V for some discussion of this assumption of linearity.

structures. Some of the results are plotted on Figures 5 and 6. The first of these demonstrates the effects of altering the gasoline and the fuel price structure upon the optimal production of fuel. (The gasoline output must always equal 100 gallons minus the yield of fuel.) Figure 6, the second of this pair, relates the total consumption of ethyl fluid to the gasoline and the fuel oil price structure.[6]

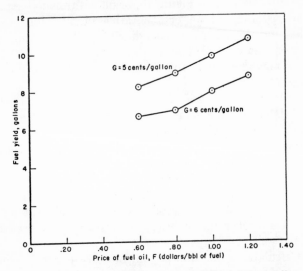

Fig. 5. Optimum Fuel Yield vs Price of Fuel Oil

Even without a formal graphical technique, the qualitative results of these charts are in accord with commonsense notions. That is, for a given value of gasoline, *as the value of fuel oil is increased*, the greater will be the optimum production of fuel, the smaller the production of gasoline and the smaller the consumption of ethyl fluid. Similarly, for a given value of fuel oil, as the price of gasoline increases, it pays to produce more gasoline, less fuel, and to consume a larger quantity of ethyl fluid. To the analyst, the advantage of having such charts as 5 and 6 lies, not

[6] The eight points on each of these diagrams are based upon Table 2, p. 43, of the Adams-Creelman study. The "wiggles" on these curves can be attributed mostly to the fact that the values of X were varied by discrete 10-gallon steps, rather than by infinitesimal increments between 30 and 80 gallons.

primarily in predicting the *direction* of response to changes in the price structure, but rather in gauging the quantitative extent of such responses. Here, for example, it turns out that the supply of gasoline and of fuel is inelastic with respect to changes in the price of fuel oil.[7] According to Figure 5, at a gasoline price of

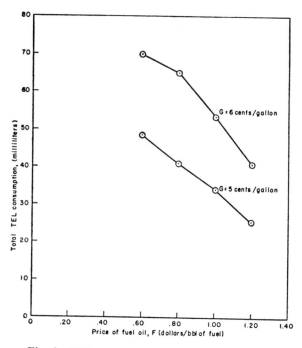

Fig. 6. TEL Consumption vs Price of Fuel Oil

6¢ per gallon, as the price of fuel oil is doubled from 1.43¢ per gallon ($0.60 per barrel) to 2.86¢ ($1.20 per barrel), the optimal production of fuel only increases by $\frac{2.1}{6.7} = 31$ percent and the production of gasoline decreases by $\frac{2.1}{93.3} = 2.3$ percent.

By contrast, it seems that the optimal yield of fuel and that the

[7] This same general result also holds true for the thermal cracking refinery discussed in Chapter VI; see part 6 particularly.

consumption of ethyl fluid are quite sensitive to changes in the gasoline price structure. For example, at a fuel oil level of 2.38¢ ($1.00 per barrel), as the price of gasoline is raised from 5¢ to 6¢ per gallon (i.e., an increase of only 20 percent), the output of fuel goes down from 9.9 to 8.0 gallons (19 percent), and the consumption of ethyl fluid rises from 34.2 to 53.3 milliliters (56 percent). Without the aid of some such calculations as these, even an individual with a well-trained intuition might not have guessed at the difference in the effects of a change in the price of gasoline and a change in the price of residual oil. The results are not at all symmetrical.

The reader is cautioned *not* to read too much into these findings. The analysis refers to only one intra-refinery process; it does not allow for the possibilities of shifting between alternative sources of crude oil, and it assumes that the refinery realizations do not decline as the volume of sales increases. If all these complexities were somehow taken into account, it is unlikely that the particular quantitative predictions would remain applicable.

6. ALGEBRAIC PROCEDURE

As a preliminary step in enlarging the scope of the study to include several operating processes, it looked essential to recast the Adams-Creelman system in algebraic form. Once the basic assumptions of the earlier work had been understood, it turned out to be quite easy to make this transition. Indeed, to any true mathematician, the quadratic approximation device employed here will seem like nothing more than an elementary calculus exercise. The fact remains that this technique makes it possible to introduce many additional degrees of freedom into the problem, and at the same time to reduce the volume of clerical work that would be needed in order to establish a series of nomograms such as Figures 4a through 4g.

In solving the basic maximizing equation (IV.4), the graphical results suggested several special mathematical features, all of which were employed in the calculus solution. First, the optimum values of X and Y did not occur at corner points, for instance at a 100-gallon reformer charge or at a zero reformer yield. Second, there were not several *isolated* optimal values for X and Y. And third, the observed solutions lay in a restricted region.

It did not seem probable, for example, that the optimal reformer yield, Y, would even be as low as 0.60.

None of these generalizations are inherent in the pure logical form of equation (IV.4). If, for the sake of argument, the price of gasoline jumped up from $0.05 to $1.00 per gallon and all other prices remained the same, it is certain that the refiner would not find it profitable to reform any of this particular raw naphtha. He could always meet the 70 octane specification by adding TEL,

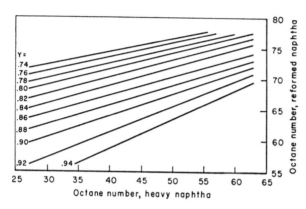

Fig. 7. Correlation between Gasoline Yield, Octane Number of Reformer Charge Stock, and Octane Number of Gasoline Product

and in this way avoid converting any $1.00 gasoline into fuel that is worth only $0.02 or $0.03 per gallon. Profit maximization would coincide with maximizing the gasoline output. Needless to say, the calculus apparatus is not intended to deal with such radical changes in the production incentives.

Starting from equation (IV.4) several new definitions were adopted in order to facilitate the calculations.

$$
\begin{aligned}
b_1 &= 100\,(G - S); \\
b_2 &= F + S - G - P; \\
b_3 &= G - S - F;
\end{aligned}
$$

$E(XY)$ = Number of milliliters of TEL required to raise   (IV.5) the total gasoline product to 70 octane C.F.R.-M.; a function of X and Y;

T = Price of TEL (cents per milliliter).

Inserting these new definitions into equation (IV.4), the following simplification occurs:

$$C = b_1 + b_2X + b_3XY - T E (X,Y). \qquad (IV.6)$$

When reading equation (IV.6), it must be borne in mind that E, the ethyl fluid requirement, is not an independent variable, but is a function of X and Y — just as in the case of the previous unknowns L_V and L_R. Fortunately, Adams and Creelman provide

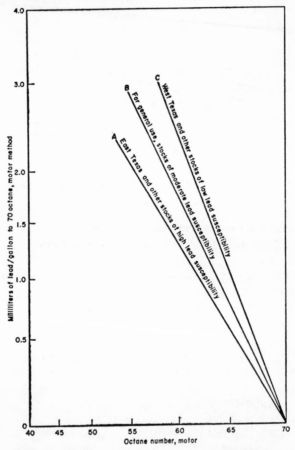

Fig. 8. TEL Concentration Requirements for 70 Octane Finished Gasoline

data that can be manipulated in such a way as to build up Table IV.2 — an array giving values of $E(X,Y)$ for various arguments of X and Y. In order to construct this table, five charts were needed — Figures 4a through 4c reproduced above and two additional ones, Figures 7 and 8.

Figures 4a and 4c provide all that is necessary in order to determine the relationship between the *single* variable X and the TEL concentration per gallon of light gasoline. The procedure here is the same as steps (1) through (3), pp. 56–57.

In addition, Figures 7 and 8 make it possible to determine the *joint* influence of the two independent variables upon the lead concentration level in the reformer gasoline product. In determining this concentration level, the chart-user must start with a given value of X in 4a, for example, 40 gallons. With this, he goes into 4b in order to estimate the octane number of the heavy naphtha, i.e., 29. From there, he moves to Figure 7. Entering with the value of 29 octane number for the heavy naphtha, and assuming a reformer yield of 0.83, he finds that the reformed gasoline will then have an octane number of 66.5. Finally, employing the 66.5 octane estimate, he inspects Figure 8, selects the line reading "B, for general use, stocks of moderate lead susceptibility," and determines that the required lead concentration level will be 0.48 ml per gallon. This result of 0.48 was obtained only by assuming arbitrary values of X and Y, respectively, 40 and 0.83. For each additional value of X or of Y, the process must be repeated — from Figures 4a to 4b to 7, and then to 8. By proceeding in this manner, and by translating the concentration levels into the total TEL requirements, it is possible to determine each entry in Table IV.2.

For convenience in interpretation, this same information is reproduced as Figure 9, a chart relating E, the total lead consumption, to X and Y, the two degrees of freedom. The chart, as might be expected, indicates that there are sharply increasing incremental lead costs with respect to both X and Y. For a given number of gallons of reformer charge, as the amount of gasoline yield is increased (i.e., a less severe reforming treatment), there is a progressively higher marginal lead requirement. Similarly, for a given degree of severity (i.e., a constant value of Y), the greater the number of gallons charged to the reformer, the smaller will be the marginal *reduction* in lead consumption.

Table IV.2. Ethyl Fluid Consumption as a Function of Reformer Charge and Gasoline Yield

	X, Reformer charge (gallons)						
Y, Proportion gasoline yield	0	30	40	50	60	70	80
0.94	300	—	—	—	—	150.21	143.14
.92	300	—	—	130.02	112.56	101.96	93.68
.90	300	—	125.88	101.80	83.40	70.17	62.08
.88	300	—	108.97	81.34	62.30	45.24	36.72
.86	300	—	95.17	65.61	43.13	25.58	15.20
.84	300	—	83.98	51.82	28.32	12.08	3.38
.82	300	—	73.61	41.93	19.10	2.07	—
.80	300	—	66.92	34.90	10.80	—	—
.78	300	107.34	61.80	28.99	4.31	—	—
.76	300	103.86	57.54	23.76	—	—	—
.74	300	100.78	54.10	20.66	—	—	—

In examining Table IV.2 and Figure 9, it is necessary to recall that these apply to only one specific naphtha, and that they would have to be altered for any change in the properties of this material. If another type of naphtha were to be considered as a candidate for reforming, it would be necessary to have a new cut point — octane relation drawn on Figure 4a, and a new TEL susceptibility curve on 4c. The user would also be required to specify which particular lead susceptibility line on Figure 8 was applicable. The basic correlation of Figure 7 remains valid regardless of variations in the type of raw naphtha material that happens to be available. Once all the changes had been established, a new set of $E(X,Y)$ curves could be constructed by following the same steps as those connected with Figure 9.

After a number of points for $E(X,Y)$ have been calculated, it is a relatively simple matter to fit a quadratic form to them. The problem is one of determining the a_i coefficients in the following equation:

$$E = a_1 + a_2X + a_3Y + a_4X^2 + a_5Y^2 + a_6XY. \quad (IV.7)$$

In order to evaluate the a_i, six combinations of X and Y were selected, points that were all within the range of previously observed optima. These six are encircled on Figure 9 (40,0.84), (50,0.80), (50,0.84), (50,0.88), (60,0.80), and (60,0.84). Inserting the corresponding value of $E(X,Y)$ into equation (IV.7) gives a system of six simultaneous linear equations, with the six a_i as

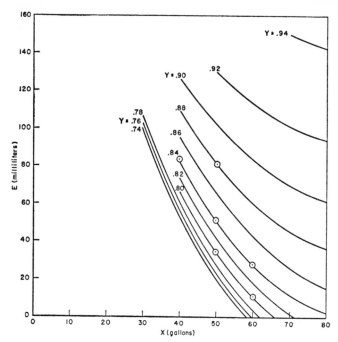

Fig. 9. Ethyl Fluid Consumption, E, as a Function of Reformer Charge, X, and Yield, Y

unknowns. The numerical values obtained for these coefficients were as follows:

$$a_1 = 2,652.90, \qquad a_4 = 0.0433,$$
$$a_2 = -8.373, \qquad a_5 = 3,937.5,$$
$$a_3 = -6,109.5, \qquad a_6 = 1.5.$$

Table IV.3, immediately following, gives selected values of $E(X,Y)$ as determined by the quadratic form (IV.7), and compares these with the corresponding points calculated graphically

Table IV.3. Comparison of Two Methods of Predicting TEL Consumption

X	Y	$E(X,Y)$ as determined graphically in Table IV.2 (ml)	$E(X,Y)$ as determined algebraically through equation IV. 7 (ml)	Ratio of algebraic $E(X,Y)$ to graphical $E(X,Y)$ (percent)
40	0.90	125.88	132.085	104.93
40	.88	108.97	112.900	103.61
40	.84	83.98	83.980	100.00
40	.80	66.92	67.660	101.11
40	.76	57.54	63.940	111.12
50	.90	101.80	100.825	99.04
50	.88	81.34	81.340	100.00
50	.86	65.61	65.005	99.08
50	.84	51.82	51.820	100.00
50	.82	41.93	41.785	99.65
50	.80	34.90	34.900	100.00
50	.78	28.99	31.165	107.50
50	.76	23.76	30.580	128.70
60	.90	83.40	78.225	93.79
60	.86	43.13	41.805	96.93
60	.84	28.32	28.320	100.00
60	.82	19.10	17.985	94.16
60	.80	10.80	10.800	100.00
60	.78	4.31	6.765	156.96
40	.84	83.98	83.98	100.00
50	.84	51.82	51.82	100.00
60	.84	28.32	28.32	100.00
70	.84	12.08	13.48	111.59
80	.84	3.38	7.30	215.98

and reproduced in Table IV.2. The equation gives an exact fit, of course, at the six special points; the two predictions of ethyl consumption check closely for points intermediate between these six; and they give a poorer fit the further away they lie from this neighborhood. In sixteen out of the twenty-one independent observations recorded here, the two methods diverge by less than 10 percent in predicting the lead requirements. Note, however, that the goodness of fit cannot be as high as that yielded by a least-squares regression.

The six a_i that have been derived do not enter directly into the final statement of the problem. Instead, six new coefficients are defined:

$$
\begin{aligned}
c_1 &= b_1 - T\, a_1 = 100(G - S) - T\, a_1; \\
c_2 &= b_2 - T\, a_2 = F + S - G - P - T\, a_2; \\
c_3 &= -T\, a_3; \\
c_4 &= -T\, a_4; \\
c_5 &= -T\, a_5; \\
c_6 &= b_3 - T\, a_6 = G - S - F - T\, a_6.
\end{aligned}
\qquad \text{(IV.8)}
$$

Inserting the six c_i coefficients into equation (IV.6), the basic maximizing equation becomes:

$$ C = c_1 + c_2 X + c_3 Y + c_4 X^2 + c_5 Y^2 + c_6 XY. \quad \text{(IV.9)} $$

In order that the refinery realization, C, be a maximum at a point (X^0, Y^0) with a proper derivative, the following conditions are necessary:[8]

$$ \frac{\partial C^\circ}{\partial X^\circ} = 0; \qquad \frac{\partial C^\circ}{\partial Y^\circ} = 0. \qquad \text{(IV.10)} $$

Differentiating (IV.9) as indicated, and solving for X° and Y°, one obtains:

$$ X^\circ = \frac{c_3 c_6 - 2 c_2 c_5}{4 c_4 c_5 - c_6{}^2}, \qquad \text{(IV.11)} $$

$$ Y^\circ = \frac{c_2 c_6 - 2 c_3 c_4}{4 c_4 c_5 - c_6{}^2}. \qquad \text{(IV.12)} $$

The solution of the entire naphtha reforming problem is stated in the two elementary relations (IV.11) and (IV.12). In any particular application, once the user determines appropriate values for the economic parameters $G,S,F,P,$ and T, and for the six a_i ethyl coefficients, the information can be inserted into the two equations, and the optimal X° and Y° may be read off directly. For purposes of comparison, the solutions were deter-

---

[8] The second-order sufficiency conditions are frequently of interest, but will not be treated here. On the significance of these conditions, see P. A. Samuelson, *Foundations of Economic Analysis* (Cambridge, Mass., 1947), esp. chapter IV, pp. 57–89.

mined through equations (IV.11) and (IV.12) for eight parameter variations, and then set alongside the corresponding ones given in Tables 1 and 2 of the Adams-Creelman report (pp. 39 and 43). The two sets of results are reproduced in Table IV.4.

Table IV.4. Comparison of Optimal Reforming Conditions — Nomogram and Algebraic Methods

Price structure (cents per gallon)		Nomogram technique, Tables 1 and 2, Adams & Creelman		Algebraic technique, equations (IV.11) and (IV.12)	
		$X°$ (percent)	$Y°$	$X°$ (percent)	$Y°$
Gasoline 5.00	Fuel				
	1.43	55	0.850	55.36	0.8451
	1.90	55	.835	55.32	.8367
	2.38	55	.820	57.58	.8275
	2.86	60	.820	60.15	.8176
6.00	1.43	50	.865	50.04	.8620
	1.90	50	.860	51.48	.8543
	2.38	50	.840	53.18	.8460
	2.86	55	.840	55.15	.8374

Upon examining Table IV.4, it will be evident that the two competing methods yield virtually identical answers for the optimum reforming conditions. There is no noticeable difference in the order of accuracy of the two approaches. If two-dimensional problems were the only ones that ever arose in practice, the chief distinguishing feature of an algebraic format would be the fact that equations (IV.11) and (IV.12) could reduce some of the labor involved in solving for the effects of any specific parameter variation. It is also true that the quadratic form is an easy one to fit, and leads to simple expressions for $X°$ and $Y°$.

Any clerical savings along these lines, however, are marginal. If this were the only issue involved, a refiner would be fully justified in clinging to his familiar nomogram devices. The true advantage of algebraic methods lies in the possibility of tackling problems with many degrees of freedom — for example, the gasoline blending problem discussed in the chapter that immediately follows. As soon as an operator wishes to consider the production of several grades of gasoline, and as soon as he wishes to take account of a large number of possible components in the blended products, he faces two alternatives. Either he makes rough-and-ready assumptions in order to reduce the problem to a small number of dimensions, or he may elect to form an algebraic model. The structure of such mathematical models need not be confined to the elementary calculus type. It is feasible to include corner-point solutions and many forms of nonlinearities within a more advanced framework. Automatic computing machines represent a logical extension of the existing graphical methods for numerical analysis.

7. SUMMARY AND CONCLUSIONS

This chapter has offered a case study in the mathematical methods currently being employed by refinery economics managers. When the analysis is confined to small sections of an integrated operation, the present nomogram techniques appear to be perfectly adequate. The only disadvantage in relying upon such methods consists of the fact that they cannot be extended to problems that involve many independent variables. At the same time, refinery economic balance calculations typically *do* involve multiple interlocking decisions. Under such circumstances, there is real danger that intuition will be substituted in place of straightforward programming calculations.

Even if it were possible to mechanize many of the existing trial-and-error solutions, there would remain ample room for the exercise of judgment and intuition in refinery scheduling. The hardest job in such a problem is the initial choice of empirical data. Unless experienced refinery people collaborate in formulating problems and in interpreting the results, a mathematical economist will only succeed in constructing elegant, but empty models of the production processes.

Chapter V

A Gasoline Blending Problem[1]

1. BASIC ECONOMIC ASSUMPTIONS

The preceding chapter has dealt with one type of intra-refinery economic balance problem — the operation of a typical kind of conversion unit. The present one concentrates upon another such intra-refinery study — the selection of an optimal gasoline blending schedule. Not until the following chapter will any attempt be made to discuss the operations of an integrated refinery.

There were two primary reasons for singling out the blending problems of the Union Oil Company of California. First, this operation is itself a complex large-scale one, grossing an amount of the order of $100 millions annually. And second, the data available for this analysis were of tolerable precision. It was agreed that there is little point in making an elaborate economic study when the basic numbers are shaky.

Together with members of the Manufacturing Economics Division of Union Oil, the problem was formulated in the following terms (see the schematic diagram, Figure 10). First, the quantities and qualities of nineteen raw gasoline stocks were taken as fixed. Together with tetraethyl lead (a purchased item),

[1] In connection with this section, a particular debt of gratitude is owed to Messrs. Reaugh, McCreery, and Norton of the Manufacturing Economics Division of the Union Oil Company of California. Without their encouragement and active help, it would not have been possible to carry through the work described here.

these serve as the inputs to the gasoline blending operation. The input materials are combined to form three basic types of products: 7600, premium grade gasoline; 76, regular grade gasoline; and fuel oil cutter stock. As a first approximation, each of these products was considered to be salable at a refinery realization that was independent of the total quantity marketed. The scheduling problem consists of assigning each raw gasoline stock among the three potential products, and also of setting the tetraethyl lead levels for the two motor fuels. This assignment has to take into account not only the absolute quantities of materials that are available, but also certain product specifications on the two finished gasolines.[2]

Each of these assumptions must be reviewed with care. Consider, for example, the matter of blending stocks. With the exception of just two of the nineteen inputs, it is assumed here that the quantities of raw gasolines available and that the quality characteristics of these materials are independent of decisions taken in the blending department. In fact, from the viewpoint of a refinery superintendent, the quantities and characteristics of all the streams are variables subject to control. By altering the choice of crudes, the reactor temperatures, the recycle ratios, and the assignment of intermediate oil streams, the central management is able to influence the volume and composition of the gasoline blending materials. This type of interlock can only be studied in a larger model that cuts across departmental lines.

A second questionable point is the assumption that additional quantities of finished refinery products may be marketed at the prevailing realizations. In the case of a major refiner, this is open to serious doubt. It is not at all obvious that a large company can put additional quantities on the market without disturbing the current price structure.[3] At the same time, no one — least of all an economist — is in a position to make a reasonable

---

[2] During the course of short time intervals, the quantities and qualities of the available blending stocks may change radically. An unforeseen shutdown of a unit can take place. A new unit may be brought on stream. A new source of crude oil may be introduced, etc. For these reasons, in order to keep blending analyses up to date, it is the current practice within the company to recalculate schedules at least once a month. The optimum allocations of raw streams among the various uses could be significantly affected if, say, a fire occurred in a large catalytic cracker.

[3] See J. S. Bain, *The Economics of the Pacific Coast Petroleum Industry* (Berkeley, Part I, 1944; Part II, 1945; and Part III, 1947).

quantitative estimate of the relationship between a company's price structure and the demand for its products.

For purposes of computation, an explicit marketing limitation could have been placed upon each finished product. In order to provide a benchmark, however, it was decided to study one basic case and then one in which a sales restriction was introduced. The comparison of these two provides an indication of the extent to which profits could be improved by overcoming the marketing barrier. The comparison also indicates the magnitude of the gap between the average refinery realization and the incremental product cost.

Corresponding to the assumption about the realization upon refinery products, it was agreed that one of the inputs, tetraethyl lead, would be considered as available without any quantity restrictions at the prevailing tank-car delivered price of the Ethyl Corporation. During wartime, TEL has been rationed among refiners, but at least during 1953 this type of restriction did not hold true. If lead were to become short again, the availability limitation would have to be considered as an explicit element in the problem. As shown in the preceding chapter, it is possible for the refiner to substitute reformed and cracked gasolines over a wide range in place of tetraethyl lead in the finished products. Although lead is employed in virtually every gallon of gasoline today, its use in motor fuel *could* be eliminated, but only at a distinct cost in dollars and cents to the refiner.

Having decided upon the two major assumptions — fixed gasoline stock availabilities and fixed refinery product realizations — the next step was to determine which of the numerous quality characteristics of the finished products would be studied explicitly. The Manufacturing Economics Division pointed out that there were many specifications that finished motor fuel must meet — three types of octane numbers; 10 percent, 30 percent, 60 percent, 90 percent, and end points of the boiling range; sulfur and gum content; vapor pressure, etc. Any of these could conceivably limit the output of either premium or regular grade gasoline. For purposes of this study, the ones that seemed to be worth the most close attention were: Research octane number, the percent sulfur content, and a volatility index. In addition, the endpoint problem was handled automatically by limiting the end points of the individual raw gasolines. During the course of the computing runs, it turned out that sulfur was not a limiting specification, but that the octane number and volatility index specifications did come into play.

In order to protect proprietary information, the exact house-brand specifications cannot be revealed here. It can, however, be said that the 7600 premium gasoline has a higher minimum octane number and volatility index, and a lower maximum permissible sulfur content and end point than the regular 76 gasoline. On both products, the maximum permissible tetraethyl fluid concentration is 3.0 ml per gallon.

Although the blending section must operate within whatever company specifications are currently effective, the reader should not suppose that these limits remain fixed over time. At any moment, "octane wars" between refiners can break out. The effect upon motorists, and upon refinery profits, is of much the same nature as an actual price war. For this reason, a separate computation was set up in order to determine the effects of a 2.5 octane number increase in the specifications of the premium gasoline.

Before proceeding to the technology of the problem, two more of the economic assumptions should be brought out — one having to do with the cutter stocks and the other with transportation costs. The raw gasolines are manufactured, and the blending operations are performed at two distinct locations — at Wilmington in the Los Angeles area and at Oleum on San Francisco Bay. Raw stocks are continually lifted by tanker for interchange between the two refineries. Union Oil estimates its tanker costs between the two points to be nominal, only a few percent of the refinery realization on premium gasoline. In the analysis of the company's gasoline blending problems, it is evidently a desirable thing to take account of transportation costs and of differences in product realizations at the two refineries. At the same time, it was easy to see that by bringing geography into the problem, the volume of computations would have increased by more than 100 percent. The storage capacity of the then available computing machine, the International Business Machine Card Programmed Calculator, would have been completely inadequate to take care of the enlarged model.[4] As a first effort, therefore, it was agreed that the geography should be left out of account. The raw gasoline stocks were considered to be available in one company-wide pool, and the finished products were valued at the company's overall realizations. (Roughly speaking, these revenue figures are of the same order of magnitude as the current spot market refinery tank-car quotations.)

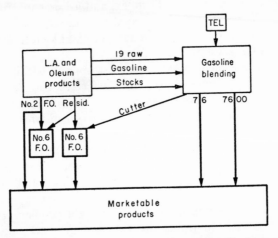

Fig. 10. Union Oil Co. Gasoline Blending Problem

Finally, there is the problem of cutter stock realizations. As indicated by Figure 10, a number of the heavy gasolines may be employed, not only for motor fuel components, but also for blending with heavy residuum into No. 6 fuel oil. The figure also shows that this heavy residuum may be converted into No. 6 oil by the alternative route of blending with No. 2 furnace oil.[5] Under the assumption that *some* heavy residuum is always going to be cut down to the viscosity specifications by blending with No. 2 oil, it is possible to establish a value for the heavy residuum,

[4] This type of limitation would *not* have existed if one of the higher-capacity electronic computers had been available, such as the IBM 701 or the Univac.

[5] In general, the cutter will not be a finished No. 2 oil, but rather some gas oil component that *could* be used for No. 2. When there are several alternative gas oils that may be employed for cutter, the problem takes on additional complexity. See Chapter VI below, pp. 140–141.

and then to go on to determine a cutter value for any particular heavy gasoline.

To give a purely hypothetical illustration of the procedure, assume the following product values:

No. 2 furnace oil $4.00 per barrel;
No. 6 fuel oil $2.00 per barrel;
Percent of No. 2 oil in No. 6 blend required for cutting back heavy residuum to No. 6 oil specifications 25 per cent;
Percent of heavy gasoline in No. 6 blend required for cutting back heavy residuum to No. 6 oil specifications 20 per cent.

If No. 2 oil is used for blending, the value of the heavy residuum, p_r, may be established through the following equation:

Value of No. 6 oil = Value of No. 2 oil + Value of residuum
component   component
$$\$2.00 = 0.25(\$4.00) + (1 - 0.25)(p_r)$$
$$p_r = \$1.333 \text{ per barrel.}$$

Next, this value for p_r is substituted in a costing equation for the heavy gasoline cutter stock. Out of this relation, there is derived p_g, the cutter value of the particular gasoline.

Value of No. 6 oil = Value of gasoline + Value of residuum
component   component
$$\$2.00 = 0.20(p_g) + (1 - 0.20)(\$1.333)$$
$$p_g = \$4.668 \text{ per barrel.}$$

Values for each of the eight possible gasoline materials were calculated in this manner. Certain treatment cost savings also had to be credited to the cutter operation. The final calculated net cutter values range from 68.2 percent to 87.4 percent of the refinery realization on premium gasoline. In effect, this is an opportunity cost calculation of the same nature as the "gasoline replacement" technique described above, pp. 14–16. Like the gasoline replacement formula, this method applies to the comparison of just two alternative refining operations. (In the one case, the alternatives are cracking versus fuel oil, and in the other, gasoline versus No. 2 fuel oil cutter materials.) In the event that the company's residuum were capable of meeting the fuel oil specifications without any additional cutter, the heavy gasoline could not be given the premium value as cutter, but would have to be assigned a lower credit.

2. THE BLENDING TECHNOLOGY

Of the three relevant gasoline properties, two — the volatility index and the percent sulfur content — give rise to straight-forward linear blending relationships. For example, in a 50-50 blend of two raw gasolines with sulfur contents of 0.10 percent and 0.30 percent, the resulting mixture will contain 0.20 percent sulfur. Similarly with the volatility index, a 50-50 blend will have an index that lies midway between those of its two component raw gasolines.

Unfortunately, the octane number of blends cannot be predicted in such a direct manner from the octane numbers of the components. Like most kinds of engineering work, the forecast of gasoline octane numbers is an art rather than an exact science. Refiners themselves necessarily make paper predictions of the octane number of proposed blends, but before marketing a product they will almost inevitably take the precaution of testing the mixture in an octane rating engine.

One phase of the problem has been investigated extensively by the refining industry — the relationship between octane numbers and the TEL concentration of a particular gasoline. Just as in the naphtha reforming problem above, it has been observed that, as the lead concentration level is increased, the octane number increases at a decreasing rate. In order to predict this relationship, there is in widespread use today an ethyl blending nomogram put forward by Hebl, Rendel, and Garton in 1939.[6] There is, fortunately, a three-parameter analytic expression that gives a close approximation to the results predicted by this blending chart. Where t represents the octane number of the leaded gasoline, and x the TEL concentration level (ml/gal),

$$t = a + bx - \frac{c}{1 + x}. \qquad (V.1)$$

[6] L. E. Hebl, T. B. Rendel, and F. L. Garton, "Ethyl Fluid Blending Chart for Motor-Method Octane Numbers," *Industrial and Engineering Chemistry*, 31:862–865 (July 1939).

For any unknown gasoline, the three positive parameters, a, b, and c, have to be determined. This may be accomplished readily by taking three observations of octane number for various lead levels, and solving the resulting simultaneous linear equations. Using equation (V.1) in this way, sets of constants were calculated for all nineteen of the Union Oil Company's raw gasolines. Testing seven lead levels in each of these nineteen cases, the equation has virtually always given a prediction that lies within ± 0.2 octane number of the one yielded by the blending chart.

No special significance should be attached to the particular form of equation (V.1). It is merely a device for enabling a computing machine to perform the same calculation as a refiner with his ethyl chart and his straight edge.

A second phase of the blending problem is more controversial than the TEL aspect. For a lead level of, say, 3.0 milliliters, and for a 50-50 mixture of two gasolines A and B, refiners frequently calculate the octane number of the blend to be the 50-50 weighted average of the octane number of gasolines A and B — each with 3.0 milliliters. The weighted average appears satisfactory for many stocks, especially when both components are of a paraffinic nature. There are, however, at least two papers publicly available — one by Eastman[7] and the other by Bogen and Nichols[8] — that call into question the straight-line average method. Both papers indicate that as the percentage of the high octane component in a binary mixture increases (both components initially at the same TEL concentration level), the octane level of the mixture may increase at a decreasing rate. In other words, the octane number of the blend tends to *exceed* the weighted average of the octane number of the two components.

Bogen and Nichols are primarily concerned with three-component mixtures, and they report only five observations at each of three TEL levels for strictly binary blends. The set of

[7] Du Bois Eastman, "Prediction of Octane Numbers and Lead Susceptibilities of Gasoline Blends," *Industrial and Engineering Chemistry*, 33:1555–1560 (December 1941).

[8] J. S. Bogen and R. M. Nichols, "Calculating the Performance of Motor Fuel Blends," *Industrial and Engineering Chemistry*, Vol. 41:2629–2635 (November 1949).

fifteen points has been plotted, and is reproduced as Figure 11. For each TEL concentration level shown on this chart, straight-line interpolation has been used between adjacent observations. In all likelihood, these straight lines also underestimate the octane number that is attainable by blending. The existing data, unfortunately, do not provide a basis for curvilinear interpolation.

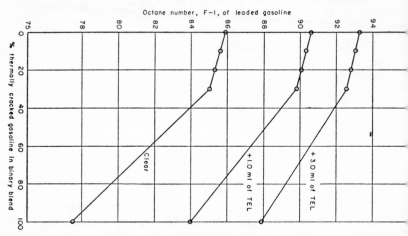

Fig. 11. Octane Number vs Percent of Thermally Cracked
Component in Binary Blend. Three TEL Levels

At first glance, it would appear to be a trivial matter whether octane numbers were calculated by drawing straight lines between two adjacent points or between two end points of a constant-TEL line. Nevertheless, a sample calculation presented in an earlier paper[9] indicates that the two methods can lead to TEL input requirements that differ by 20 percent.

It is evidently worthwhile for a refiner to incur considerable expense in testing gasoline blends, and in attempting to predict the occurrence of curvilinear octane relationships. Nevertheless, at the present stage, the Union Oil Company did not have such blending data available, and it was necessary to revert back to the more conventional straight-line interpolation procedure for

[9] A. S. Manne, "Concave Programming for Gasoline Blends" (P-383, The RAND Corporation, Santa Monica, California, April 23, 1953).

calculating the octane numbers of blends.[10] The information currently at hand made it possible for the analysis to take account of nonlinear TEL relationships, but not of the nonlinear gasoline blending features.

The TEL input is one of the elements in the Union Oil Company problem that involved a departure from a strictly

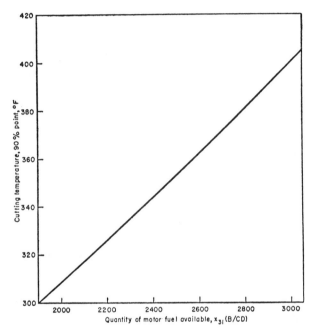

Fig. 12a. Cutting Temperature vs x_{31}

linear system. In addition, the cutting temperatures on two of the gasoline distillation units introduced nonlinear effects. For purposes of blending, just as in the Kellogg Company naphtha reforming problem, the cut point on each of these units is an

[10] In going back to the simple straight-line octane interpolation procedure, this paper follows the model proposed by A. Charnes, W. W. Cooper, and B. Mellon, "Blending Aviation Gasolines — A Study in Programming Interdependent Activities in an Integrated Oil Company," *Econometrica*, vol. 20, no. 2 (April 1952), pp. 135–159. But unlike the aviation gasoline system, the present one includes ethyl fluid concentration levels as a specific variable with a curvilinear effect.

independent variable that is subject to control. The cutting temperature influences, not only the absolute quantity of material that is available for motor fuel blending, but also the quality of this component. Figures 12a through 12c contain the necessary data for one of these two distillation units — the one producing gasoline stock #31.

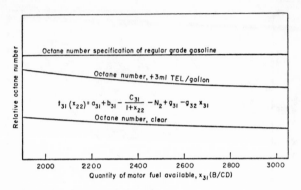

Fig. 12b. Clear and Leaded Octane Numbers vs x_{31}

According to Figure 12a, as the cutting temperature moves from 300° to 400°F, the amount of material available for motor fuel blending rises steadily from 1,900 B/CD to 3,050 B/CD. (For each additional barrel of motor fuel obtained, there is approximately a one-barrel decrease in the amount of gasoline cutter stock available.)

The charts of Figures 12b and 12c contain the original Union Oil Company curves, and also indicate the form of the equations that were fitted to these data. As the make of motor fuel increases, Figure 12b indicates a virtually linear decline in the octane number, both clear and with 3 ml of TEL per gallon. Also, as the make of motor fuel increases, there is a steady linear decrease in the volatility index, and a quadratically increasing sulfur content.[11]

---

[11] Although when distilling this particular stock the sulfur content increases quadratically, on the other material (gasoline #36) the sulfur content only changes linearly.

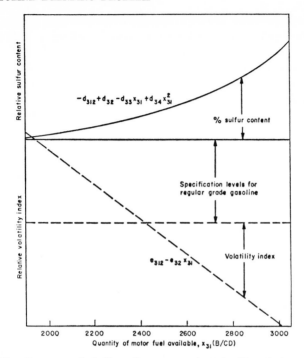

Fig. 12c. Percent of Sulfur Content and Volatility Index vs x_{31}

3. THE MATHEMATICAL MODEL

3.1 THE BASIC FORM

Kuhn and Tucker have proved the following "equivalence theorem": "Let the functions $f_1(x)$, . . . $f_m(x)$, $g(x)$ be concave as well as differentiable for $x \geq 0$. Then x^0 is a solution of the maximum problem if, and only if, x^0 and some u^0 give a solution of the saddle value problem for $\phi(x,u) \equiv g(x) + u'Fx$." [12]

---

[12] The expression $u'Fx$ is shorthand for the following: $u_1 f_1(x) + u_2 f_2(x) + \cdots + u_m f_m(x)$. The u_i are nonnegative Lagrangean multipliers, and correspond closely to the economist's notion of "shadow prices." The symbol (x) represents a vector, a set of variables.

H. W. Kuhn and A. W. Tucker, "Nonlinear Programming," *Proceedings of the Second Berkeley Symposium on Mathematical Statistics and Probability*, ed. J. Neyman (Berkeley, 1951), p. 486.

In other words, they have established that the solution to a concave constrained-maximum problem is equivalent to a certain minimax saddlepoint. At the suggestion of Harry Markowitz, the gasoline blending model was set up in the appropriate minimax form. An iterative digital technique was then devised for solving the problem.[13]

From a rigorous mathematical standpoint, the present computing procedure is an unsatisfactory one, but the method does lead to solutions that should be sufficiently accurate from the standpoint of refinery managers. Convergence takes place without an inordinate amount of computing time. All calculations were performed on a machine that is readily available throughout the United States — an IBM Card Programmed Calculator, Model 2.

Dr. Martin Beckman has objected to the application of the Kuhn-Tucker theorem in this particular instance. He has made the point that for a function of many variables, $f(x)$, to be truly concave, it is necessary but *not* sufficient for that function to be concave in *each variable taken separately*. It has been possible to devise examples that satisfy all conditions of the gasoline blending problem — that is, concave in each variable taken separately — but which nevertheless violate the strict Kuhn-Tucker conditions.[14] The counterexamples give rise to multiple isolated maxima. Without the concavity property in the various func-

---

[13] Markowitz has experimented with an electric analogue solution for this class of applications. The present blending model involving nineteen gasoline stocks would clearly swamp the capacity of any of the present generation of analogue machines.

[14] The following example is not derived from any Union Oil Company data, and serves only for illustrative purposes. Consider two blending stocks, labeled A and B with the following characteristics:

	Stock A	Stock B
Octane number, clear	74.0	90.0
Octane number, with 3.0 ml of TEL	90.0	94.8
Number of ml/gal required for 90 octane leaded gasoline	3.0	0

Assuming straight line interpolation at constant lead levels for the two stocks, and employing the Hebl-Rendel-Garton blending chart, it will be found that a 50-50 mixture would require 1.6, and *not* 1.5, ml/gal in order to reach the 90.0 octane level. This clearly constitutes a violation of the concavity assumption, but a refiner will also recognize that the example requires an unusual configuration of octane numbers and lead susceptibilities.

tions, it is possible to adopt blending schedules that are locally optimal, but which do not, in fact, represent the best of all solutions available.

The problem cannot be dismissed lightly, but I suspect that in most particular applications such local extrema can be detected by the investigator. The difficulties arising out of nonconcavity are not unique to the programming method described here. The same argument would apply if the TEL levels were taken as parameters, and a separate linear programming problem then run off for each of many *combinations* of the two. This difficulty would come up in the case of any trial-and-error method that consisted of changing one variable at a time, and observing the effects upon the payoff. And finally, the failure of the proper curvature conditions could frustrate optimizations by way of a Lange-Lerner shadow price market mechanism.

3.2 DEFINITION OF INDEPENDENT VARIABLES

It now remains to set up the formal relationships corresponding to the Kuhn-Tucker payoff function, $g(x)$, and the set of re-restraints, $f_i(x)$. Table V.1 presents schematically the 23 independent non-negative variables, x_j, which must be selected in order to determine the blending schedule. The whole set of 23 will be termed the vector (x).

There are nineteen raw gasoline stocks, labeled 31, 36, 41, ... , 80, as indicated by the row headings. In each instance, we are given q_j, the total quantity of this material available for blending. The problem consists of assigning various amounts of the raw gasolines among the three alternative uses — (1) 7600, premium grade motor fuel, (2) 76, regular grade motor fuel, and (3) cutter stock. In addition, it is necessary to select x_{21} and x_{22}, respectively, the ethyl fluid concentrations in premium and regular grade gasoline.

The computations were *not* set up with the full generality of considering $3(19) = 57$ individual possible gasoline components. Instead, for purposes of economizing on machine capacity, it was agreed to make advance guesses at which of the elements would turn out to have zero values. (Wherever a component was pre-assigned in this way, there is a zero entry in the corresponding box of Table V.1.) It was also agreed that it would not be

Table V.1. Gasoline Blending Problem — Definition of Various Quantities

Identification No.	Stocks		Quantity of stock j assigned to three products (B/CD)		
	Quantities available		(1) 7600 premium motor fuel	(2) 76 regular motor fuel	(3) Cutter stock
	q_i	B/CD			
31	q_{31}	3,050	0	x_{31}	$(q_{31} - x_{31})$
36	q_{36}	2,060	0	x_{36}	$(q_{36} - x_{36})$
41	q_{41}	1,000	$(q_{41} - x_{41})$	x_{41}	0
42	q_{42}	2,910	x_{45}	x_{42}	$(q_{42} - x_{42} - x_{45})$
43	q_{43}	3,460	x_{46}	x_{43}	$(q_{43} - x_{43} - x_{46})$
44	q_{44}	1,730	0	x_{44}	$(q_{44} - x_{44})$
51	q_{51}	110	$(q_{51} - x_{51})$	x_{51}	0
52	q_{52}	1,240	x_{55}	x_{52}	$(q_{52} - x_{52} - x_{55})$
53	q_{53}	860	x_{56}	x_{53}	$(q_{53} - x_{53} - x_{56})$
54	q_{54}	490	0	x_{54}	$(q_{54} - x_{54})$
60	q_{60}	6,150	$(q_{60} - x_{60})$	x_{60}	0
61	q_{61}	3,430	$(q_{61} - x_{61})$	x_{61}	0
62	q_{62}	1,460	$(q_{62} - x_{62})$	x_{62}	0
63	q_{63}	3,200	$(q_{63} - x_{63})$	x_{63}	0
64	q_{64}	1,680	$(q_{64} - x_{64})$	x_{64}	0
65	q_{65}	11,230	$(q_{65} - x_{65})$	x_{65}	0
66	q_{66}	7,440	$(q_{66} - x_{66})$	x_{66}	0
70	q_{70}	1,700	$(q_{70} - x_{70}) \equiv q_{70}$	$x_{70} \equiv 0$	0
80	q_{80}	8,710	$(q_{80} - x_{80}) \equiv 0$	$x_{80} \equiv q_{80}$	0
Totals	—	61,910	z_1	z_2	—
TEL concentration (ml/gal)			x_{21}	x_{22}	—

economical to throw away any of these particular gasoline stocks.[15] In this manner, it was possible to reduce the number of independent gasoline components from 57 down to 21.

[15] Even if a material could not be employed directly as cutter or as regular grade gasoline, under the 1953 price structure it would always pay to degrade some finished 7600 gasoline for purposes of blending the particular material up to the specifications for regular grade gasoline.

The Lagrangean multipliers, u_i, taken directly from the basic calculations, enabled us to check up on these particular assumptions at a later stage in the analysis. Fortunately, the initial guesses turned out to be correct ones. Despite the good luck for the particular sets of parameters studied, these assumptions were an unsatisfactory short cut, and can be justified only on grounds of computational convenience.

Table V.1 always refers to the quantity of stock j in regular grade gasoline as x_j.[16] Even in the case of stocks #31 and #36, where the cutting temperature can equally well be regarded as the independent variable, this same convention was employed. (On Figure 12a note the one-to-one relationship between the cut point and the quantity of motor fuel available.)

To simplify further reference, two new *dependent* variables are also defined:

$$z_1 = \sum_j (q_j - x_j) + \sum_k x_k = \text{total production rate (B/CD) for premium grade gasoline}$$

$$(j = 41, 51, 60\text{-}66, 70; k = 45, 46, 55, 56). \qquad (V.2)$$

$$z_2 = \sum_j x_j \qquad = \text{total production rate (B/CD) for regular grade gasoline}$$

$$(j = 31, 36, 41\text{-}44, 51\text{-}54, 60\text{-}66, 80). \qquad (V.3)$$

3.3 DEFINITION OF THE PAYOFF FUNCTION AND RESTRAINTS

The payoff, $g(x)$, is determined by crediting the gross refinery realization from the sale of premium and regular grade gasoline, and then charging up the ethyl fluid costs and the *loss* of potential cutter credits associated with the particular blending schedule:

$$g(x) = (p_1 - p_{20} x_{21})z_1 + (p_2 - p_{20} x_{22})z_2$$
$$- p_{42} x_{45} - p_{43} x_{46} - p_{52} x_{55} - p_{53} x_{56}$$
$$- \sum_j p_j x_j$$
$$(j = 31, 36, 42, 43, 44, 52, 53, 54). \qquad (V.4)$$

---

[16] For symmetry in the subsequent work, there are two additional constants that are written *as if* they were independent variables. The term x_{70} is identically zero, and x_{80} identically equals q_{80}.

In equation (V.4), the various unit costs and prices are defined as follows:

p_1 = refinery realization on 7600, premium gasoline (dollars per barrel),

p_2 = refinery realization on 76, regular grade gasoline (dollars per barrel),   (V.5)

p_{20} = cost of ethyl fluid (dollars per 42 ml),

p_j = unit cutter credit on stock j (dollars per barrel)

$$(j = 31, 36, 42, 43, 44, 52, 53, 54).$$

The blending schedule (x) must not call for the production of negative amounts of any motor fuel or cutter component, and so there are certain upper and lower limits imposed upon the various x_j. In addition, the schedule must not require TEL concentration levels exceeding 3.00 ml per gallon. And because of the flash-point specification on finished No. 6 oil, neither of the potential cutter components may fall below q_{31}^0 and q_{36}^0, respectively. Thus,

$$\begin{aligned}
0 &\leq x_{21} \leq 3.0, \\
0 &\leq x_{22} \leq 3.0, \\
q_{31}^0 &\leq x_{31} \leq q_{31}, \\
q_{36}^0 &\leq x_{36} \leq q_{36}, \\
0 &\leq x_i \leq q_i, \\
0 &\leq x_j
\end{aligned}$$

(V.6)

$$(i = 41, 44, 51, 54, 60\text{--}66; \quad j = 42, 43, 45, 46, 52, 53, 55, 56).$$

For stocks #42, 43, 52, and 53, the quantities of individual cutter stocks, $f_i(x)$, must not be negative;

$$\begin{aligned}
f_1(x) &= q_{42} - x_{42} - x_{45} \geq 0; \\
f_2(x) &= q_{43} - x_{43} - x_{46} \geq 0; \\
f_3(x) &= q_{52} - x_{52} - x_{55} \geq 0; \\
f_4(x) &= q_{53} - x_{53} - x_{56} \geq 0.
\end{aligned}$$

(V.7)

The six remaining equations are connected with the three specifications on each of the two motor fuel products. The functions $f_5(x)$ and $f_6(x)$ relate, respectively, to the octane number of premium and regular grade motor fuel; $f_7(x)$ and $f_8(x)$ to the sulfur content; and $f_9(x)$ and $f_{10}(x)$ to the volatility index. These relationships are all defined in such a way that the particular specification will be satisfied if, and only if, the function $f_i(x) \geq 0$. To simplify notation, the constants in the equations do not refer to the absolute octane number, sulfur content, or volatility index of a particular raw gasoline, but rather to the *differences* between these values and those of the specifications for a particular motor fuel product.

In the case of the octane ratings, for example, equation (V.1) indicates the relationship between the octane number of a particular blending stock, and the TEL concentration in a finished product. A new function, $t_{jk}(x_{2k})$ is defined in order to indicate the number of octane points by which stock j exceeds the specifications for product k, when x_{2k} milliliters of lead are added to stock j. The variable t_{jk} is evidently a function of x_{2k}, and is determined through the following relation:

$$t_{jk}(x_{2k}) = a_j + b_j\, x_{2j} - \frac{c_j}{1 + x_{2k}} - N_k \quad (k = 1,2). \qquad (V.8)$$

The terms a_j, b_j, and c_j represent the lead susceptibility constants for stock j, and correspond to the three constants in equation (V.1). The parameter N_k stands for the minimum octane number requirement connected with gasoline blend k.

In order to ensure meeting octane number specifications on the premium grade gasoline, the following condition must then apply:

$$f_5(x) = \sum_j t_{j1}\, (q_j - x_j) + \sum_i t_{i1}\, x_i \geq 0$$
$$(j = 41, 51, 60\text{--}66, 70;\ i = 45, 46, 55, 56). \qquad (V.9)$$

Similarly, to make sure that the regular grade gasoline blend will pass its octane number specification, we must have:

$$f_6(x) = \sum_j t_{j2} x_j + x_{31}(g_{31} - g_{32} x_{31})$$
$$+ x_{36}(g_{36} - g_{37} x_{36}) \geq 0$$
$$(j = 31, 36, 41\text{--}44, 51\text{--}54, 60\text{--}66, 80). \qquad (V.10)$$

The functions in parentheses that follow the variables x_{31} and x_{36} are inserted in order to reflect the fact that with these two gasolines, as the cutting temperature changes, both the quantity and the octane number of the motor fuel component will be affected. (See Figure 12b). This feature evidently leads to negative square terms involving x_{31} and x_{36}.

The same two stocks give rise to quadratic expressions in connection with the sulfur and volatility specifications for regular grade gasoline, but since these components were assumed to be absent from the premium gasoline blend, those equations remain completely linear. For the sulfur specification on the premium grade gasoline, the condition reads

$$f_7(x) = \sum_j d_{j1} (q_j - x_j) + \sum_k d_{k1} x_k \geq 0$$

$$(j = 41, 51, 60\text{--}66, 70; \; k = 45, 46, 55, 56). \quad \text{(V.11)}$$

The constant term d_{j1} represents the algebraic excess of the maximum allowable sulfur content of 7600 gasoline over the amount actually in stock j. There is a similar set of constants associated with the regular grade gasoline — d_{j2}, the difference between the allowable sulfur content in the 76 grade product and the content in stock j. The following relationship controls the sulfur level in the finished regular grade product:

$$f_8(x) = \sum_j d_{j2} x_j - x_{31}(d_{32} - d_{33}x_{31} + d_{34}x_{31}^2)$$

$$- x_{36}(d_{37}x_{36}) \geq 0$$

$$(j = 31, 36, 41\text{--}44, 51\text{--}54, 60\text{--}66, 80). \quad \text{(V.12)}$$

The terms in parentheses following x_{31} and x_{36} arise out of the cutting temperature relationship. (See Figure 12c.) Note that over the entire range between q_{31}^0 and q_{31}, the terms involving x_{31} bring about a strictly concave function. That is, the following partial second derivative is negative over this particular interval —

$$\frac{\partial^2 f_8(x)}{\partial x_{31}^2} = 2d_{33} - 6d_{34}x_{31} < 0. \quad \text{(V.13)}$$

The functions $f_9(x)$ and $f_{10}(x)$ do the job, respectively, of governing the volatility of the premium and regular grade gasoline.

$$f_9(x) = \sum_j e_{j1} (q_j - x_j) + \sum_k e_{k1} x_k \geq 0$$

$$(j = 41, 51, 60\text{--}66, 70; \ k = 45, 46, 55, 56). \quad \text{(V.14)}$$

$$f_{10}(x) = \sum_j e_{j2} x_j - e_{32}x_{31}^2 - e_{37}x_{36}^2 \geq 0$$

$$(j = 31, 36, 41\text{--}44, 51\text{--}54, 60\text{--}66, 80). \quad \text{(V.15)}$$

Here again, because of the cutting temperature-quality relationship, there are square terms connected with x_{31} and x_{36}.

The gasoline blending model has now been reduced to a constrained-maximum mathematical form. The problem consists of choosing a vector (x) that will maximize $g(x)$, subject to the side conditions that $f_i(x) \geq 0$, and also subject to certain upper and lower limits on the individual components, x_j. The payoff function $g(x)$ is given by equation (V.4). The upper and lower limits on individual variables are stated in (V.6), and the 10 restraining inequalities, $f_i(x)$, in (V.7), (V.9), (V.10), (V.11), (V.12), (V.14), and (V.15).

In this form, the problem *almost* fits the Kuhn-Tucker conditions. Both the payoff function and the restraints are concave in each variable taken separately.[17] Furthermore, eight of the ten restraints are concave over the whole space — $f_1(x) \ldots f_4(x)$, and $f_7(x) \ldots f_{10}(x)$. The payoff function $g(x)$ and the two octane number restraints — $f_5(x)$ and $f_6(x)$ — do not strictly satisfy the full concavity properties. It was decided, nevertheless, to perform the computations as if these also satisfied the rigorous Kuhn-Tucker conditions. Under this approach, there is always the danger that a local optimum will not turn out to be the *maximum maximorum*. A number of alternative computing schemes were examined, but all of these suffered from precisely the same defect.

---

[17] Kuhn and Tucker give the following definition of concavity: "A function is *concave* if linear interpolation between its values at any two points of definition yields a value not greater than its actual value at the point of interpolation." "Nonlinear Programming," p. 481. The function is said to be concave in each variable separately if the two points of definition are always chosen so that only one of the components of the vector (x) is altered. A function may be concave in each variable separately *without* being concave over the whole space of points (x).

The primary justification for going ahead in this nonrigorous fashion was that the computational results could always be checked against the experience of the refinery engineers. It was felt that these individuals would certainly be able to detect any errors of a gross nature.

4. THE NUMERICAL ANALYSIS

4.1 OUTLINE OF COMPUTATIONAL PROCEDURE

Following the assumption that this gasoline blending model fits into the format of the Kuhn-Tucker theorem, it is necessary to define the Lagrangean function $\phi(x,u)$:

$$\phi(x,u) = g(x) + \sum_{i=1}^{10} u_i f_i(x). \qquad (V.16)$$

The constrained-maximum problem will be solved if, and only if, there is a saddlepoint solution to the function, $\phi(x,u)$. The computational procedure is an iterative one — at each step t, converting a vector $x(t)$, $u(t)$ into a new vector $x(t+1)$ and $u(t+1)$. The solutions observed have all tended toward a saddlepoint, but I can give no strict proof of the necessity of this convergence. The new vector generated is never exactly "efficient," nor is it "attainable" in the sense of Koopmans.[18] The payoff does not increase monotonically as in each successive step of Dantzig's simplex procedure for linear programming.[19] Despite these apparent shortcomings, the algorithm has given useful answers in the cases examined to date.

The solution must be started off from some initial point $x(0)$, $u(0)$. In principle, this may be any arbitrary nonnegative vector. In practice, though, it is possible to effect a considerable reduction in computing time if a good initial position is selected. At each step, first the (u) vector is determined, and then the (x) vector. For the former, the basic iteration consists of two steps, and for the latter, of four steps. The problem is one of determining

[18] T. C. Koopmans, "Analysis of Production as an Efficient Combination of Activities," *Activity Analysis of Production and Allocation*, chap. III, p. 79.

[19] G. B. Dantzig, "Maximization of a Linear Function of Variables Subject to Linear Inequalities," *Activity Analysis of Production and Allocation*, chapter XXI, pp. 339–347.

$\Delta u_i \equiv u_i(t+1) - u_i(t)$, and $\Delta x_j \equiv x_j(t+1) - x_j(t)$. The procedure appears cumbersome, but only six minutes of CPC machine time are actually required for generating a whole set of Δu_i and Δx_j. (The required total number of steps has varied between 50 and 80.) For the Δu_i, the procedure is as follows:

1. If $\dfrac{\partial \phi(t)}{\partial u_i(t)} \geq 0$, then the "candidate" $\langle \Delta u_i \rangle = -k_i$.

   If $\dfrac{\partial \phi(t)}{\partial u_i(t)} < 0$, then the "candidate" $\langle \Delta u_i \rangle = k_i$.

Note: k_i is an arbitrary positive constant.

2. If $u_i(t) + \langle \Delta u_i \rangle < 0$, then $\Delta u_i = -u_i(t)$.
   If $u_i(t) + \langle \Delta u_i \rangle \geq 0$, then $\Delta u_i = \langle \Delta u_i \rangle(t)$.

For the Δx_j, the procedure becomes:[20]

1. If, for any i, $f_i(t) < 0$, and if $\dfrac{\partial \phi(t)}{\partial x_j(t)} \cdot \dfrac{\partial f_i(t)}{\partial x_j(t)} < 0$, then $\Delta x_j = 0$;

proceed to evaluate Δx_{j+1}. Otherwise, proceed to step 2.

2. If $\dfrac{\partial \phi(t)}{\partial x_j(t)} \geq 0$, then the "candidate" $\langle \Delta x_j \rangle = k_j$.

   If $\dfrac{\partial \phi(t)}{\partial x_j(t)} < 0$, then the "candidate" $\langle \Delta x_j \rangle = -k_j$.

Note: k_j is an arbitrary positive constant.

3. a. If $j = 31$ or 36, and if $x_j(t) + \langle \Delta x_j \rangle < q_j^0$, then $\Delta x_j = x_j(t) - q_j^0$; proceed to evaluate Δx_{j+1}.
   If $j = 31$ or 36, and if $x_j(t) + \langle \Delta x_j \rangle \geq q_j^0$, then proceed to step 4.

   b. If $j \neq 31$ or 36, and if $x_j(t) + \langle \Delta x_j \rangle < 0$, then $\Delta x_j = -x_j(t)$; proceed to evaluate Δx_{j+1}.
   If $j \neq 31$ or 36, and if $x_j(t) + \langle \Delta x_j \rangle \geq 0$, then proceed to step 4.

4. a. If $j = 42, 43, 45, 46, 52, 53, 55$ or 56, $\Delta x_j = \langle \Delta x_j \rangle$; proceed to evaluate Δx_{j+1}.

[20] The individual Δx_j are determined in ascending order of their respective indices, j. To simplify the notation, Δx_{j+1} is used to indicate the next Δx_j that is to be determined. For example, Δx_{36} follows Δx_{31} in sequence, and for $j = 31$, is indicated by Δx_{j+1}.

b. If $j = 21$ or 22, calculate $3.00 - x_j(t) - \langle \Delta x_j \rangle$.

If this expression is nonnegative, $\Delta x_j = \langle \Delta x_j \rangle$. Otherwise $\Delta x_j = 3.00 - x_j(t)$. Proceed next to evaluate Δx_{j+1}.

c. If $j = 41, 44, 51, 54$, or $60, 61, \ldots, 66$, calculate $q_j - x_j(t) - \langle \Delta x_j \rangle$. If this expression is nonnegative, $\Delta x_j = \langle \Delta x_j \rangle$. Otherwise $\Delta x_j = q_j - x_j(t)$. Proceed next to evaluate Δx_{j+1}.

For the Δu_i, the explanation of this ritual is straightforward. Step (1) tells us to decrease u_i by an arbitrary amount k_i if, for a "small" change in u_i *alone*, the effect will be to decrease $\phi(x,u)$. Similarly, there is to be an increase in u_i if, for a small increase, ϕ would decrease. Step (2) prevents u_i from becoming negative.

For the Δx_j, the justification is more roundabout. [Steps 3 and 4 are the obvious ones — respectively, lower and upper limits on the individual x_j dictated by (V.6).] As with the Δu_i, the sign of $\partial \phi(t)/\partial x_j(t)$ is primarily the criterion that determines whether to take a positive or a negative step.[21] Preliminary small-scale calculations indicated, however, that the unqualified "direction of ascent (descent) rule" would lead to major oscillations rather than to convergence at equilibrium values. For that reason, rule 1, a relaxation-type principle, was inserted. Translated into English, the statement reads, "If one of the restraints is being violated, and if the direction-of-ascent rule would lead to an even greater violation of this restraint, then the particular x_j should not be changed." The relaxation rule is hardly elegant from the viewpoint of pure mathematics, and there are undoubtedly other possibilities for achieving the same effect.

At the present time, despite their plausibility, there is a serious dilemma concerning the logic of these computational techniques. On the one hand, I have been unable to prove the necessity of convergence for either the relaxation or the unmodified procedure. On the other, I have been unable to construct a counterexample for a case where the k_i and the k_j may be made arbitrarily small. Readily granting these apparent objections, the fact remains that the relaxation technique has given useful results in a series of computing runs.

[21] The reader should note that ϕ is being minimized with respect to (u), and maximized with respect to (x).

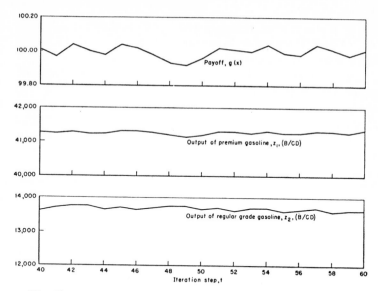

Fig. 13a. Run Number 1, Payoff and Total Gasoline Output

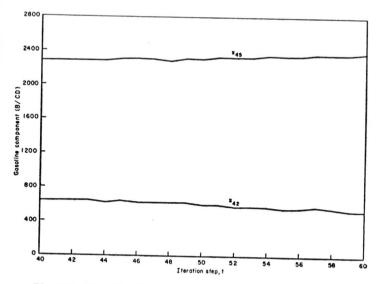

Fig. 13b. Run Number 1, Gasoline Components x_{42} and x_{45}

4.2 DETAILS OF THE NUMERICAL ANALYSIS

Three sets of calculations have been performed. The first is to be considered the basic case — taking as parameters the 1953 refinery realizations, product specifications, and raw material availabilities. The second run was concerned with a radical change in the price structure — an 11.6 percent drop in the value of premium grade gasoline. In the third, the price structure was taken to be the same as in the basic case, but a 2.5 octane point

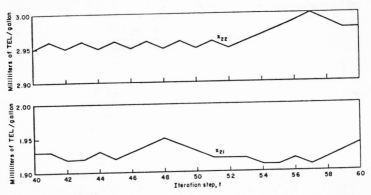

Fig. 13c. Run Number 1, TEL Concentration Levels x_{21} and x_{22}

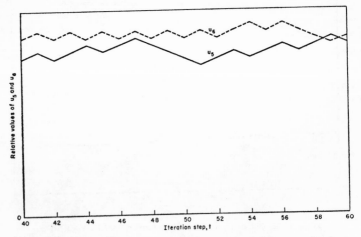

Fig. 13d. Run Number 1, Langrangean Multipliers, u_5 and u_6

increase was imposed upon the premium product. For each of these departures from the basic conditions, the optimal blending schedule becomes considerably altered.

In order to convey some indication of the computational technique employed here, Figures 13a through 13e chart the values of certain of the variables during the last twenty steps of run number 1, i.e., between $t = 40$ and $t = 60$. According to Figure 13a, the payoff and the aggregate make of premium and regular grade gasoline had become stabilized, and "rippled" steadily

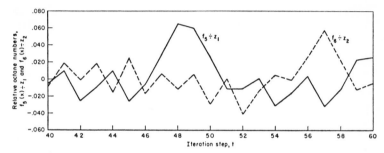

Fig. 13e. Run Number 1, Relative Octane Numbers $f_5(x) \div z_1$ and $f_6(x) \div z_2$

about certain equilibrium levels. Virtually all of the independent variables exhibited similar rippling around an equilibrium point, but there were exceptions. The worst drifting occurred in the case of x_{42} and x_{45}, respectively, the quantities of stock #42 in regular and premium grade gasoline. Their time paths are shown in Figure 13b. The more typical pattern was exhibited by the two TEL concentration variables, x_{21} and x_{22}, and is reproduced in Figure 13c. Two of the Lagrangean multipliers, u_5 and u_6, connected with the two octane number specifications, are shown in Figure 13d. The octane number functions themselves, $f_5(x)$ and $f_6(x)$, are given in Figure 13e.

Before cutting off any individual computation run, it was possible to perform an additional check upon the stability of the solution. According to the Kuhn-Tucker theorem, in order for the vector $(x)^0$ to be a saddle value solution, the following condition must hold for all x_j^0:[22]

[22] "Nonlinear Programming," p. 482.

$$\frac{\partial \phi^0}{\partial x_j^0} \leq 0, \quad x_j^0 \frac{\partial \phi^0}{\partial x_j^0} = 0, \quad x_j^0 \geq 0. \tag{V.17}$$

In the present case, where the x_j are bounded from above as well as below, this condition on the equilibrium x_j^0 becomes:[23]

a. If $\partial \phi^0 / \partial x_j^0 \leq 0$, then $x_j^0 \dfrac{\partial \phi^0}{\partial x_j^0} = 0$, and $x_j^0 \geq 0$;

b. If $\partial \phi^0 / \partial x_j^0 > 0$, then $x_j^0 = q_j$. $\qquad$ (V.18)

Using condition (V.18), it is possible to perform a rough check on the precision of the solution. For those x_j that are at their lower limits, $\partial \phi^0 / \partial x_j^0 \leq 0$. For those that are at their upper limits, $\partial \phi^0 / \partial x_j^0 \geq 0$. And for those that are neither at their upper nor at their lower limits, $\partial \phi^0 / \partial x_j^0 = 0$. In this latter case, it is not to be expected that the partial derivative will be literally at a zero level. Rather, since this is an approximative method of solution, the derivative should be in the *neighborhood* of zero.

Table V.2 contains the actual values for the $x_j(60)$ and for the normalized values of the partial derivatives $\partial \phi(60) / \partial x_j(60)$. All of those variables that are at their lower limits exhibit negative partial derivatives, as is to be expected. On the other hand, the single variable that lies at its upper limit, x_{54}, does not have a partial derivative of the proper sign. This derivative is slightly negative, and in fact, during the sequence leading up to $t = 60$, $x_{54}(t)$ had oscillated between 470 B/CD and 490. There appears to be a coincidence between the upper limit on this particular variable, and the point at which $\partial \phi / \partial x_{54}$ vanishes.

In the case of those variables that lie neither at their upper nor their lower limits, the maximum absolute deviation from zero amounts to 2.35 percent of the value of one barrel of premium grade gasoline. This deviation occurred in the case of x_{45}, the variable shown in Figure 13b that had been slowly drifting upward. It did not seem worthwhile to continue the iterations in order to refine upon this one estimate.

[23] In the case of (V.18a), where $j = 31$ or 36, the lower limits q_j^0 must be interpreted as equivalent to zero. For (V.18b), where $j = 21$ or 22, the upper limits of 3.00 ml are to be regarded as the appropriate q_j. For those x_j on which there is no specific imposed upper limit — that is, $j = 42, 43, 45, 46, 52, 53, 55$, and 56 — the condition (V.18b) does not apply.

Table V.2. Comparison of Partial Derivatives for $t = 60$, Run No. 1

j	Lower limit on x_j	$x_j(60)$	Upper limit on x_j	$\dfrac{\partial\phi(60)}{\partial x_j(60)}$ (percentage of realization on 1 barrel of premium gasoline)	Adjusted optimal x_j
21	0	1.94	3.00	1.57 percentage of cost of 1 ml/gal in premium gasoline	1.92
22	0	2.98	3.00	−0.61 percentage of cost of 1 ml/gal in regular gasoline	2.96
31	1,900	1,900	3,050	−22.54	1,900
36	1,320	1,320	2,060	−0.34	1,320
41	0	0	1,000	−5.58	0
42	0	540	2,910	−0.88	620
43	0	400	3,460	−0.40	380
44	0	0	1,730	−8.62	0
45	0	2,390	2,910	2.35	2,290
46	0	850	3,460	1.09	845
51	0	0	110	−4.41	0
52	0	0	1,240	−3.41	0
53	0	280	860	−0.05	260
54	0	490	490	−0.08	480
55	0	100	1,240	0.26	120
56	0	600	860	1.27	600
60	0	0	6,150	−6.90	0
61	0	0	3,430	−5.98	0
62	0	0	1,460	−9.82	0
63	0	0	3,200	−8.55	0
64	0	0	1,680	−9.10	0
65	0	0	11,230	−7.94	0
66	0	0	7,440	−7.35	0

The blending schedule selected at step 60 is neither literally "efficient" nor "attainable." The payoff $g(x)$ is below the level recorded for step 45, and for this reason, the final blending schedule was taken over from that earlier step. Certain minor corrections were made by hand in order to bring about a feasible and efficient solution. The resulting vector is listed in the final column under the heading "Adjusted optimal x_j."

5. RESULTS OF THE NUMERICAL ANALYSIS

5.1 INTERPRETATION OF RESULTS — BLENDING PROGRAMS

The operations schedules derived from the three computation runs are presented in Table V.3. For reference purposes, the first

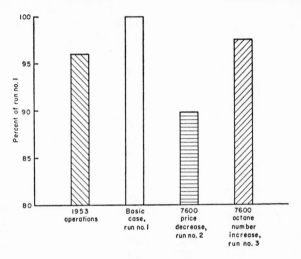

Fig. 14a. Total Payoff, $g(x)$, Four Cases

three columns indicate the program that was effective at the beginning of the year 1953. The other sets give the results for run number 1, the basic case; run number 2, a decrease of 11.58 percent in the realization on premium grade gasoline; and run number 3, an increase of 2.5 octane numbers in the specification for the premium product.

The outstanding features of these different schedules are summarized in bar charts, Figures 14a through 14d. Of all four situations, the output of 7600 gasoline is largest in the case of run number 1, and smallest under "1953 operations." Just the reverse is true for the regular grade gasoline. Its production is largest under the 1953 operations plan, and smallest in the basic

Table V.3. Results of Computations

Stocks	Quantities available (B/CD)	Quantity of stock assigned to three products (B/CD)											
		1953 operations			Basic case run no. 1			7600 price decrease run no. 2			7600 octane no. increase run no. 3		
Identification number		7600	76	Cutter stock	7600	76	Cutter stock	7600	76	Cutter stock	7600	76	Cutter stock
31	3,050	—	2,260	790	—	1,900	1,150	—	1,900	1,150	—	1,900	1,150
32	2,060	—	1,890	170	—	1,320	740	—	2,060	0	—	1,320	740
41	1,000	500	500	0	1,000	0	0	1,000	0	0	1,000	0	0
42	2,910	1,455	1,455	0	2,290	620	0	1,175	1,735	0	2,690	220	885
43	3,460	0	3,460	0	845	380	2,235	170	3,260	30	1,050	1,525	1,730
44	1,730	—	1,730	—	—	—	1,730	—	1,040	690	—	—	—
51	110	55	55	0	110	0	0	70	40	0	110	0	1,240
52	1,240	620	620	0	120	0	1,120	600	640	0	0	0	100
53	860	0	860	0	600	260	0	120	675	0	760	470	20
54	490	—	490	—	—	480	10	—	490	65	—	650	—
61	6,150	3,000	3,150	—	6,150	0	—	3,030	3,120	—	5,500	3,430	—
62	3,430	0	3,430	—	3,430	0	—	200	3,230	—	0	—	—
63	3,230	0	1,460	—	3,200	0	—	840	620	—	1,460	—	—
64	1,680	0	3,200	—	1,680	0	—	160	3,040	—	3,200	—	—
65	11,230	8,720	1,680	—	11,230	0	—	9,720	1,520	—	1,680	—	—
66	7,440	6,860	2,510	—	7,440	0	—	7,440	1,510	—	11,230	—	—
70	1,700	1,700	580	—	1,700	0	—	1,700	0	—	7,440	—	—
80	8,710	—	8,710	—	—	8,710	—	—	8,710	—	1,700	8,710	—
Totals	61,910	22,910	38,040	960	41,255	13,670	6,985	26,385	33,590	1,935	37,820	18,225	5,865
TEL concentration (ml/gal)		1.13	1.05	—	1.92	2.96	—	1.10	1.50	—	3.00	2.60	—
Total consumption of TEL (liters per day)			2,765			5,026			3,335			6,755	
Payoff, g(x), percent of total for run no. 1, basic case			96.01			100.00			89.84			97.46	
Refinery realization on motor gasoline, percent of 1953 realization on 7600 product		100.00	82.64	—	100.00	82.64	—	88.42	82.64	—	100.00	82.64	—
Octane rating of motor gasoline product, number of points in excess of 1953 housebrand specifications		0	0	—	0	0	—	0	0	—	2.5	0	—

case. Going along with this considerable shift in the general production strategy, there is an *apparently* small shift in payoff, for the basic case would yield a return to the company of only 3.99

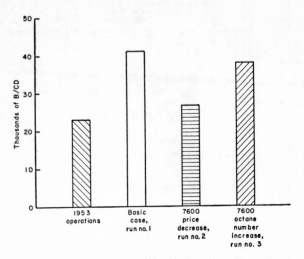

Fig. 14b. Total Output of 7600 Gasoline, Four Cases
(thousands of B/CD)

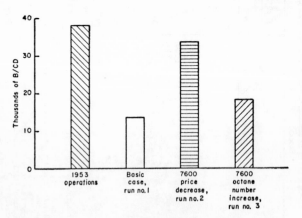

Fig. 14c. Total Output of 76 Gasoline, Four Cases
(thousands of B/CD)

percentage points in excess of that shown in actual practice.[24] At first glance this figure of 3.99 does not seem impressive, but nevertheless, during the course of a single year's operation a differential of this magnitude amounts to several millions of dollars! From the company's standpoint, it would look eminently worthwhile to shift over to the blending schedule derived from run number 1.

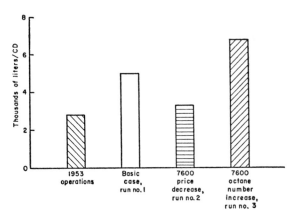

Fig. 14d. Total Consumption of TEL, Four Cases
(thousands of liters/CD)

Unhappily, the life of the operations analyst does not run a simple course. From the very beginning of the study, it had been acknowledged that the company could not realistically expect to sell unlimited quantities of premium gasoline at the prevailing refinery realization. At the same time, it had not been fully appreciated how large an increase in the production of this item might be in order. (The output in the one case is almost double that of the other.) For this reason, it seemed desirable to incorporate an additional formal restriction into the problem, and to require that the total production of premium product remain below some upper limit, L_1. In this way, it would be possible to determine how large an increase in profits is attainable — even

[24] In order to avoid revealing proprietary information, no absolute dollar figures are presented here. Both the unit and the total refinery realizations are stated in terms of index numbers.

under the limitation of *current* sales of the one product. By varying this upper limit parameter, the analyst could trace out the effects of alternative sales volumes upon the total payoff.

Rather than revise the computing setup in order to impose a direct marketing restriction of this nature, it was a more economical alternative to calculate a second point along the company's "supply curve" for premium gasoline. That is, for run number 2 all parameters were maintained at the same level as in run number 1, except for the unit realization upon 7600 gasoline. This was dropped to 88.42 percent of its previous level. The change had the effect of reducing the differential between premium and regular grade gasoline to less than half of its previous amount. This alteration in the price relations had a result analogous to that of a direct quantity limitation upon the make of premium gasoline. The optimal output of this product was reduced from 41,255 B/CD down to 26,385, while the total of regular grade 76 gasoline went up from 13,670 B/CD to 33,590. In this second case, not only premium gasoline but also cutter stock production declines. It now becomes profitable to incorporate most of the heavy gasolines into motor fuel rather than to blend them into residual fuel oil.

The results of run number 2 are equivalent to those that would be derived by employing an outright sales limitation of 26,385 B/CD. The unit price imputed to 7600 — 88.42 percent — indicates the marginal opportunity cost of this product. In the jargon of economists this value is termed the "shadow price" of the premium product. Evidently, there is a marked gap here between the shadow price and the average realization upon the product, a matter of considerable interest to A. P. Lerner and his followers.[25] This group has stressed the fact that such discrepancies are symptomatic of market restraints. At the same time, *in themselves*, these discrepancies constitute no proof of either the absence or the presence of workable competition.[26] From the viewpoint of internal management, they do provide

[25] See especially A. P. Lerner, "The Concept of Monopoly and the Measurement of Monopoly Power," *Review of Economic Studies*, 1:157–175 (June 1934).

[26] For a discussion of this concept, see J. M. Clark, "Toward a Concept of Workable Competition," *American Economic Review*, vol. 30, no. 2 (June 1940), pp. 241–256.

the company with an indication of the gross gains that might be achieved by reorienting sales policy.

One of the favorite avenues for sales rivalry in the gasoline industry consists of alterations in the specifications for the house-brand products. These modifications occasionally take the form of changes in the volatility index, but more typically center about the octane ratings. For this reason, it was of particular interest to explore the effects of an octane number increase upon the optimal blending schedule and upon the payoff. In run number 3, it was assumed that the specification of the premium gasoline had been increased by 2.5 octane points — a plausible jump in this requirement. Referring back to the bar charts, Figures 14a through 14d, the reader can compare the results with the basic case: a 2.54 percent decrease in total payoff, a moderate decrease in the output of premium 7600 fuel, and an increase in the output of 76 gasoline. The make of cutter stock remains the same magnitude in both cases.

The outstanding difference between this situation and the others consists of the high ethyl fluid consumption. It amounts to 6,755 liters per day here — about a one-third increase as against 5,026 liters per day in the basic case. The TEL concentration level is directly up at the prescribed limit of 3.00 ml per gallon for the premium grade product, but despite this fact, it is only at 2.60 ml for regular gasoline. Like the results of the reformer analysis in Chapter IV, these computations do not yield much comfort to those who would put faith in an often-quoted rule of thumb, that the optimum blending policy is to assign the maximum allowable TEL concentration to each gallon of product. Only in making premium grade gasoline in run number 3, and in making the regular product in run number 1 was the 3.00 ml limitation approached. In all other cases, it paid to employ less than this concentration.

Not only is the ethyl fluid consumption closely connected with the octane specification of the two motor fuels. It is also sensitive to a variation in the price structure for premium grade gasoline. In run number 2, the 7600 product commands a low shadow price, and by comparison with the basic case, there is accordingly a low consumption of ethyl fluid. Runs numbers 1 and 3 indicate that even during the ordinary peacetime course of market conditions, there is a high degree of flexibility possible in the use of this item. During wartime emergencies, it follows that there are substantial opportunities for cutting down on the use of the material in motor fuel so as to release TEL for aviation blends.

An interesting sidelight to the motorist: under the schedule set up in the basic case, there is actually a lower TEL concentration in the 7600 "ethyl" gasoline than in the regular grade product. This does *not* mean that the consumer fails to get his money's worth on the premium product. The ethyl fluid concentration by itself is meaningless as an indicator of gasoline quality. In order to determine the knock rating, it is also necessary to specify the composition of the product — the relative amount of components derived from straight-run gasoline, from catalytic cracking, from thermal reforming, and so on.

5.2 INTERPRETATION OF RESULTS — LAGRANGEAN MULTIPLIERS

Because of the need for withholding cost and revenue figures, it is not possible to indicate the precise numerical values for the ten Lagrangean multipliers, u_i. Nevertheless, the general nature of several results may be mentioned. Six of the Lagrangean multipliers turned out to be positive in all three runs: u_1, the material balance for stock #42; u_4, the material balance for stock #53; u_5, the octane number of 7600 gasoline; u_6, the octane number of 76 gasoline; u_9, the volatility index of 7600; and u_{10}, the volatility index of 76. Since the Lagrangean multipliers were always positive in these six cases, the restraint conditions were always satisfied exactly. It never paid to give away octane numbers or volatility index points, and it never paid to employ stock #42 or #53 as cutter.[27]

A rather different general result held true for u_7 and u_8 — the "shadow prices" associated with the sulfur content of the two finished gasolines. These two u_i consistently turned out to be zero, for with the blending stocks currently available, the company as a whole maintained a comfortable position with respect to meeting its sulfur specifications. This result had not been foreseen in advance, and it turned out to be fortunate that the re-

[27] Table V.3 shows certain quantities of stock #53 being used as cutter in the case of runs nos. 2 and 3. This is nonoptimal, but is a natural consequence of the approximate type of solution.

quirement was phrased in terms of a maximum sulfur content rather than an exact one. Had it been specified as an exact requirement, a distinctly nonoptimal program might have been generated.

Concerning the two material balance equations associated with stocks #43 and #52, no general rule can be laid down. Both u_2 and u_3 turned out to have zero values in runs numbers 1 and 3, but achieved positive levels in the case of run number 2. That is, for an optimal solution under the parameters of the second run, neither of these stocks should be employed as cutter material.

The u_i are of interest, not only in indicating whether or not the restraint conditions were satisfied as exact equalities, but also as a method for evaluating the use of additional blending material in either of the two motor fuel end products. These variables establish a scale for determining the dollar bonus or the penalty that should be attached to *small incremental* quantities of a material that either exceeds or falls short of the finished product specifications. The first use of these Lagrangean multipliers was to test the earlier assumptions as to which blending stocks could be excluded from consideration as components for 7600 and which others could be ruled out as components in the 76 gasoline product. Under the existing configuration of blending stocks and market conditions, these assumptions did check out.

It is expected that the Lagrangeans will also be of use during the interval before the company is in a position to schedule the blending problem simultaneously with the intra-refinery cracking and conversion operations. The company's Los Angeles refinery, for example, has a polyformer unit that has economic features similar to the thermal reformer discussed in Chapter IV. As with the Kellogg Company's reformer, there is a continuing problem of selecting the optimum operating conditions. The higher the reactor temperature, the better will be the octane rating of the polyformer gasoline, but the lower will be the quantity yielded. The outstanding difference between the two situations is the fact that the one reformer was considered as an isolated piece of conversion equipment, and that the other forms just a single component within an integrated company. In this second case, because of the possibilities for blending, there is no simple linkage between the market realization on the end products and the operations of the one unit. Ideally, the company-wide optimization should be treated as a single calculation, but it is doubtful whether this will be achieved within the foreseeable future. Falling short of this ideal, it would nevertheless be a useful step to attack the polyformer unit problem through the shadow prices derived from the blending calculation.[28]

5.3 COMPANY POLICY IMPLICATIONS

This much is already evident from the computations: that it would have paid for Union Oil to forego several millions per annum if a sales strategy could have been devised for disposing of an increased volume of premium gasoline. For this reason, the problem was taken up with the company's sales service manager. A number of alternatives were discussed with him: price reductions, quality improvements, and an advertising campaign. He agreed that all of these steps could conceivably increase the sales of premium grade gasoline, and that the potential increase in revenues might exceed the extra costs that were entailed. He was particularly concerned with the difficulties of forecasting the response of consumers to such changes in strategy, but did not bring up the parallel question of how the company's immediate rivals might react.

Rather than risk a major departure in price policy, he indicated that it would be useful for Union Oil to explore the possibility of altering its retailer gasoline discount structure. On any sales in excess of certain goals, the service station operators might be allowed an additional discount from the tank wagon price. By readjusting the product quotas and the sales bonuses, he believed that it would be feasible for the retail operators to shift sales in the direction of 7600 and away from 76 gasoline. This policy is one that need not immediately provoke retaliation from the company's major rivals, and it does not depend upon a large-scale publicity campaign. There is the added advantage that the

[28] For a discussion of the potentialities for decentralized decision-making through the use of shadow prices, see Oskar Lange, *On the Economic Theory of Socialism* (Minneapolis, 1938), and also A. P. Lerner, *Economics of Control* (New York, 1944). Also see C. J. Hitch, "Sub-optimization in Operations Problems," *Journal of the Operations Research Society of America*, vol. I, no. 3 (May 1953), pp. 87–99; and A. W. Marshall, "A Mathematical Note on Sub-optimization," *Journal of the Operations Research Society of America*, vol. I, no. 3 (May 1953), pp. 100–102.

strategy can be altered at short notice without making consumers aware of a deliberate attempt to influence their choices.

Even aside from major departures in company-wide sales policy, the gasoline blending model yielded one result that was expected to produce an immediate reduction in cost. In all three computing runs, it turned out to be optimal to cut motor gasoline component #31 at the lowest possible end point, that is, at 300° rather than at the previous level of 332°F. This change in operating practices could be effected without making any radical alteration in the end-item product mix. At the same time, the improvement in payoff that would result from this single step amounts to a sum which, in less than two days, could justify the entire computing machine expense involved in the study.

6. SUMMARY AND CONCLUSIONS

Employing the assumption of fixed availabilities of raw blending materials and of fixed realizations upon finished products, the Union Oil Company's gasoline blending operations were formulated in terms of a constrained-maximum problem.

The outstanding result of the numerical analysis was the fact that it would be worth several millions of dollars per annum if the company were able to shift sales so as to sell additional quantities of premium grade gasoline. As a corollary of this, it follows that the company has considerable latitude in its ability to shift production between the two grades of motor fuel. Marketing problems aside, the ideal pattern of production is quite sensitive to the price structure that prevails.

A second result of the calculations was the finding that the company could increase the octane specification of its premium product, and that even after such a change, the optimal production rate of this one item would remain in excess of current sales.

Neither of these results leads to the implication that the company "ought" to change its price policy or its product specifications. Major changes of this nature cannot be evaluated unless the company's executives also take some account of the response of consumers and of other petroleum refiners. Readily granting the uncertainty that must attend any such conjectures, the formal mathematical treatment seems like a promising approach to the blending problem. It did not turn up with solutions that are absolutely unobtainable by the customary trial-and-error techniques. Rather, it brought to light numerous assumptions that had previously lain hidden, and holds out the possibility for closer coördination between the manufacturing operations and the activities of other branches of the corporation. It is particularly important for the sales group to have quantitative information available on the possibilities for shifting between various classes of refined products, and for changing the specifications of those products. It is equally significant for them to know the cost implications of such changes.

Experience within the Union Oil Company suggests that mathematical models have considerable utility in dealing with complex production problems. This study also indicates that economists and businessmen alike are a long distance from having any satisfactory theory of how a market will respond to a change in the company's sales policies. In the absence of such knowledge, it is dangerous to place too much reliance on formal techniques. This same danger exists, though, with back-of-the-envelope calculations. In none of these matters can a mathematical model make important business decisions by itself. The most that can be claimed is that it will summarize the volume of information that executives have to act upon.

Chapter VI

Cracking, Recycling, and Blending — An Integrated Refinery Problem

1. A PREVIEW

Now that several individual elements of the scheduling problem have been studied in some detail, it is possible to assemble a few of these pieces into a more inclusive pattern. The analysis presented here is intended to cover the operations of a hypothetical thermal cracking refinery. Since this plant includes facilities for recycling the various cracked distillate oils, and since the gasoline and fuel oil items may each be blended in numerous ways, there is a substantial amount of interproduct flexibility. Under these circumstances, the methods of linear programming promise to furnish a useful tool for the job of operations scheduling.

The basic engineering assumptions supporting this study were originally developed by Emanuel Singer of the Shell Development Company.[1] At the time his study was first prepared (January 1953) it was generally believed that the recycling problem necessitated the use of nonlinear mathematics. However, through an ingenious observation of Kenneth Arrow, it turned out to be possible to convert this original system into one that fully satisfied the conditions of linear programming.

In order to acquaint the reader with the ancestry of the study, the first sections are devoted to reproducing Singer's original version. Following this, it is shown how the problem was converted into a straightforward linear programming type. The remainder of the chapter is occupied with the solution of the basic case, and with three variations on this case. The first variant deals with the effects of an increase in the refinery realization upon the Number 6 bunker fuel oil product. The second variant consists of determining the maximum output of Number 1 fuel oil consistent with several alternative requirement levels for cracked gasoline. And as a final step, the optimum construction program and product-mix is calculated for an all-new thermal refinery. Here it was possible to study the results of varying the size of the over-all capital budget. This portion of the analysis deserves more than academic interest. Through a departure from the refinery layout initially proposed by Singer, it appears possible to effect a substantial increase in the earnings of the plant, and yet stay within the same investment budget.

2. OPERATING CONDITIONS AND PRODUCT YIELDS FOR A SIMPLIFIED REFINERY

2.1 INTRODUCTION

During the sections that follow, there will be presented a series of equations that are intended to relate the details of an operating program to the over-all profitability of an idealized refinery. The plant consists of a crude-distillation column and associated steam strippers, a thermal cracking and fractionating system, and facilities for the blending and preparation of salable products.

---

[1] The numbers used for yield structure, operating costs, and capital costs were derived from published sources, and hence do not represent the actual operations of any Shell refinery. I am indebted to Robert Dorfman for having brought me into contact with Emanuel Singer's work.

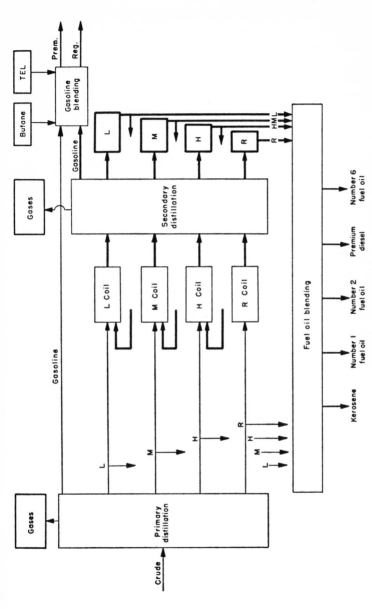

Fig. 15. Schematic Diagram of Thermal Cracking Refinery

(See Figure 15.) The thermal cracking section comprises four separate coils — one for each of three distillate streams and one for the residue. The streams emerging from these four coils enter the secondary distillation column at the appropriate places, and are fractionated into gas, gasoline, three distillates, and a residue. The boiling range of each of these materials is identical with that of the straight-run materials from the primary crude column. The end items that may be produced by this refinery are

1. Premium gasoline (88 octane number, Research),
2. Regular grade gasoline (82 octane number, Research),
3. Propane — propylene gases (C3 cut),
4. Butane — butylene gases (C4 cut),
5. Kerosene,
6. No. 1 distillate fuel oil,
7. No. 2 distillate fuel oil,
8. Premium diesel fuel (55 cetane number),
9. No. 6 bunker fuel oil,
10. Dry fuel gases (C2 and lighter).

The limited nature of the apparatus available within the refinery imposes certain restrictions on the usefulness of the intermediate product streams. As one example, in order to produce the 88 octane number gasoline without exceeding a maximum of 3.0 ml of lead per gallon, it is required that no more than 3 percent straight-run gasoline be put into this blend. If polymerization or reforming facilities were available, the straight-run component could be increased considerably.

This plant represents a relatively simple case, for it contains no equipment used specifically in the production of lubricating oils, asphalt, or any other specialty. Furthermore, it contains no facilities for catalytic cracking, polymerization, alkylation, or naphtha reforming. To this extent the problem is unrealistic; nevertheless, it furnishes a group of equations which might be typical of the operation of a more complex setup.

2.2 PROCESS DISCUSSION

The feed consists of a 38° API gravity Mid-Continent crude. The properties and amounts of the various materials present in this crude are listed in Tables VI.1 and VI.2. These cuts are obtained by distillation in a crude column. In all cases, the separations are performed at standard, pre-selected temperature levels, and the only variables to be considered are those of cracking, recycling, and blending in the subsequent processes. Since the straight-run streams are to be maintained at fixed rates of

Table VI.1. Cuts from 38.0° API Gravity Crude Oil, Mid-Continent (Mixed Base)

Cut	10 percent volume distillation temperature (°F)	End point (°F)	Percent by weight of crude	Percent by weight, cumulative
C2 and lighter	—	—	0.02	0.02
C3 gases	—	—	0.22	0.24
C4 gases	—	—	1.46	1.70
Gasoline	150	400	24.8	26.5
Light distillate	400	520	14.6	41.1
Medium distillate	510	630	9.5	50.6
Heavy distillate	620	720	12.3	62.9
Residue	—	—	37.1	100.0

flow, there is no need to consider the expenses involved in the production of these materials (heat, steam, pumping, etc.), but it is necessary to consider the cost of distillation and other processing of the effluents from the thermal cracking units.

The cracking section consists of four coils and a secondary fractionating column. There are provisions for the recycle of any portion of the three cracked distillate cuts. The overhead stream from the secondary column consists of gasoline, gases, and steam used for stripping. These tops go to a partial condenser where gasoline and steam are condensed. The gasoline is decanted, and a portion is sent back as reflux to the column. The gases then go to an absorption system, where the C3 and C4 cuts are separated, leaving a fuel gas of C2's and lighter. It is assumed that the crack-per-pass for each of the three distillate cracking coils is 0.3, and that the crack-per-pass (no recycle) for the residue is 0.5. The yields from these thermal cracking coils are given in Table VI.3, and the properties of each of the streams in Table VI.2.

Table VI.2. Product Streams, Before Blending

Material	10% volume distillation temperature (°F)	End point (°F)	Specific gravity	Octane number of gasolines and cetane number of distillates	Viscosity At 100°F		At 122°F	
					centistokes	μ_i	centistokes	μ_i
Gasoline								
straight-run	150	400	0.735	58 ON, clear	—	—	—	—
cracked	150	400	.755	75 ON, clear	—	—	—	—
Light distillate								
straight-run	400	520	.816	53 CN	1.73	−0.422	1.44	−0.5214
cracked	400	520	.840	35 CN	1.75	−.419	1.45	−.4919
Medium distillate								
straight-run	510	630	.830	58 CN	3.43	−.215	2.6	−.2965
cracked	510	630	.865	35 CN	3.37	−.220	2.65	−.2853
Heavy distillate								
straight-run	620	720	.855	61 CN	7.4	−.044	5.1	−.1215
cracked	620	720	.916	40 CN	11.2	.030	7.5	−.0417
Residue								
straight-run	—	—	.944	—	—	—	88	.2895
cracked	—	—	1.022	—	—	—	3,300	.5464

Table VI.3. Yields of Conversion Products, Single-Pass Cracking
(percentage by weight of oil converted)

| | Straight-run charge stocks | | | |
Yields	Light	Medium	Heavy	Residue (not recycled)
Gas	15	15	15	5
Gasoline	76	67	67	34
Light cycle oil	—	11	9	19
Medium cycle oil	4	—	5	21
Heavy cycle oil	3	4	—	21
Residue	2	3	4	—
Total	100	100	100	100
Total converted, percent by weight of charge	30	30	30	50

| | Cycle oil charge stocks | | |
Yields	Light	Medium	Heavy
Gas	18	16	16
Gasoline	51	48	48
Light cycle oil	—	8	8
Medium cycle oil	13	—	12
Heavy cycle oil	10	14	—
Residue	8	14	16
Total	100	100	100
Total converted, percent by weight of charge	30	30	30

There are capacity limitations on each of the parts of the refinery, and these must not be exceeded. Supposedly, there are no bottlenecks in the equipment for blending, nor in the facilities for liquefaction of the C3 and C4 cuts. All the pumps are steam-driven, and there is a high pressure steam supply available from the refinery so that low pressure steam from the exhaust of these pumps may be used for processing. In this way, all steam costs may be chalked up against process steam without introducing too large an error. For simplicity, no interstream heat exchange is considered.

Table VI.4. Product Streams, After Blending

| End product | Permissible components | Specifications | | Maximum viscosity | | | | Market price (cents per gallon) |
| | | Maximum specific gravity | Minimum octane number of gasolines and cetane number of distillates | At 100°F | | At 122°F | | |
				centi-stokes	μ_{spec}	centi-stokes	μ_{spec}	
1. Premium gasoline	SR and cracked gasoline, C4 gases, TEL	—	88 ON	—	—	—	—	11.56
2. Regular gasoline	SR and cracked gasoline, C4 gases, TEL	—	82 ON	—	—	—	—	10.44
3. Propane-propylene gases	C3 cut	—	—	—	—	—	—	4.00
4. Butane-butylene gases	C4 cut	—	—	—	—	—	—	9.00
5. Kerosene	SR light distillate	—	—	—	—	—	—	8.88
6. No. 1 fuel oil	SR and cracked light and medium distillates	0.850	—	1.9	−0.400	—	—	8.44
7. No. 2 fuel oil	SR and cracked light, medium, and heavy distillates	.882	40 CN	4.3	−.160	—	—	7.88
8. Premium diesel fuel	SR and cracked light, medium, and heavy distillates	.840	55 CN	2.6	−.290	—	—	8.63
9. No. 6 fuel oil	SR and cracked light, medium, and heavy distillates; SR and cracked residues	1.014	—	—	—	375	0.411	2.26
10. Dry fuel gases	C2 cut	—	—	—	—	—	—	0.2 cents per pound

In the blending operation, the various intermediate oils are combined to form the end products. Certain specifications must be met for each of the seven liquid products, and these are all stated in Table VI.4.

2.3 CALCULATION OF CRACKING AND RECYCLING YIELDS.

In the schematic diagram, Figure 15, it will be noted that for each pound of crude entering the crude fractionation column, a fixed proportion, S_i, of each of the straight-run streams leaves. At this point, it is necessary to make an arbitrary choice. That is, a certain proportion of each of the three distillate streams and of the residue may be sent to thermal cracking. The label F_i $(i = L,M,H,R)$ is employed in order to indicate the proportion of each straight-run stream which is used for cracking stock. The four F_i variables are each subject to the refiner's control. It is understood that the gases, S_G, may be made directly into end products, and that S_{Gaso}, the straight-run gasoline, is sent to gasoline blending.

This leaves as straight-run materials available for fuel oil blending: $(1 - F_L) S_L$, $(1 - F_M) S_M$, $(1 - F_H) S_H$, and $(1 - F_R) S_R$. The remainder of each of the four crackable streams goes to its appropriate coil along with any of the recycled materials: $R_L Z_L$, $R_M Z_M$, and $R_H Z_H$, respectively. The term Z_i refers to the *gross* number of pounds of stream i leaving the secondary column, per pound of crude distilled. The amounts of the *net* offtake from the fractionating-stripping system are designated C_G, C_{Gaso}, C_L, C_M, C_H, and C_R. With these definitions, the yield calculation may be performed in the following manner:

A. The first arbitrary choice is that of F_i $(i = L,M,H,R)$.

B. The next arbitrary choice is that of R_i $(i = L,M,H)$.

C. Because of the complications introduced by the recycle of fractions of light, medium, and heavy distillates, it is required that the following three linear equations be solved for Z_i, the gross production of the various distillate streams $(i = L,M,H)$:

$$Z_L = 0.1022F_L + 0.00314F_M + 0.00332F_H + 0.0352F_R \\ + 0.7R_L Z_L + 0.024R_M Z_M + 0.024R_H Z_H. \qquad \text{(VI.1)}$$

$$Z_M = 0.00175F_L + 0.0665F_M + 0.00185F_H + 0.039F_R \\ + 0.039R_L Z_L + 0.7R_M Z_M + 0.036R_H Z_H. \qquad \text{(VI.2)}$$

$$Z_H = 0.00131F_L + 0.00114F_M + 0.0861F_H + 0.0390F_R$$
$$+ 0.03R_LZ_L + 0.042R_MZ_M + 0.7R_HZ_H. \qquad \text{(VI.3)}$$

D. The *net* effluent of material from the secondary column and strippers is given by the following equations.

$$C_L = Z_L (1 - R_L). \qquad \text{(VI.4)}$$

$$C_M = Z_M (1 - R_M). \qquad \text{(VI.5)}$$

$$C_H = Z_H (1 - R_H). \qquad \text{(VI.6)}$$

$$C_G = 0.00657F_L + 0.00428F_M + 0.00554F_H$$
$$+ 0.00928F_R + 0.054R_LZ_L + 0.048R_MZ_M$$
$$+ 0.048R_HZ_H. \qquad \text{(VI.7)}$$

$$C_{\text{Gaso}} = 0.0333F_L + 0.0191F_M + 0.0247F_H + 0.0631F_R$$
$$+ 0.1503R_LZ_L + 0.144R_MZ_M + 0.144R_HZ_H. \qquad \text{(VI.8)}$$

$$C_R = 0.0009F_L + 0.0009F_M + 0.0015F_H + 0.185F_R$$
$$+ 0.024R_LZ_L + 0.042R_MZ_M + 0.048R_HZ_H. \qquad \text{(VI.9)}$$

2.4 CALCULATION OF END PRODUCT BLENDS

Various straight-run and cracked streams are blended together in order to meet the individual product specifications:

2.4.1 *Gasolines.* There are two grades of marketable gasoline derived from blends of straight-run gasoline, cracked gasoline, butane, and tetraethyl lead fluid. The specifications to meet are: maximum Reid vapor pressure for both gasolines, 10 pounds per square inch, and minimum octane numbers of 82 and 88, Research, respectively.

Since butane is available to this refiner at a lower cost than gasoline, it pays him to blend in the material up to the maximum Reid vapor pressure of 10 pounds per square inch. It is estimated that 10 percent by volume of the butanes in the blended gasoline will result in this maximum pressure.

The Research octane number specification is also considered to be one that should be met exactly. From Figure 16, it can be seen that the greater the proportion of straight-run gasoline in the final blend, the greater will be the TEL requirement for meeting the particular octane level. From this, it follows that there is always room to economize on ethyl fluid, and that it does not pay this refiner to "give away" octane numbers.

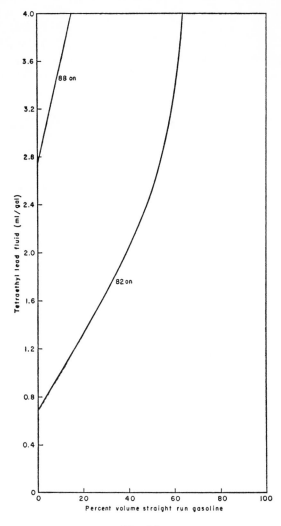

Fig. 16.

2.4.2 *Kerosene*. Salable kerosene must consist entirely of straight-run light distillate. This oil may be marketed directly as kerosene without further treatment.

2.4.3 *Number 1 distillate fuel oil*. The product is to be made from a blend of straight-run and cracked light and medium distillates. The specifications to meet are a maximum viscosity of 1.9 centistokes at 100°F and a maximum specific gravity of 0.850.

The gravity of the blend is given by the following equation:

$$\text{S.G. of blend} = \sum_i (\text{volume fraction of oil } i)(\text{S.G.})_i \quad \text{(VI.10)}$$

The viscosity of the blend may be determined as follows:

$$\mu \text{ of blend} = \sum_i (\text{volume fraction of oil } i)(\mu_i) \quad \text{(VI.11)}$$

In equation (VI.11), the viscosity blending number $\mu_i = \log \log (v + 0.6)$, where v represents the kinematic viscosity in centistokes. Since the specification viscosity is 1.9 centistokes at 100°F, $\mu_{\text{spec}} = \log \log (1.9 + 0.6) = -0.400$. The pertinent values of the viscosity and μ_i for all potential component oils are given in Table VI.2.

2.4.4 *Number 2 distillate fuel oil*. This fuel is composed of a blend of straight-run and cracked light, medium, and heavy distillates. The specifications are as follows: viscosity less than 4.3 centistokes at 100°F; a maximum specific gravity of 0.882; a minimum cetane number of 40; a flash point greater than 100°F; and an ASTM distillation end point at a maximum of 675°F. The gravity and viscosity of the blends obey equations (VI.10) and (VI.11). The cetane number is given by equation (VI.12):

$$\text{CN of blend} = \sum_i (\text{volume fraction of oil } i)(\text{C.N.})_i \quad \text{(VI.12)}$$

The flash-point and ASTM endpoint specifications may be satisfied by restricting the light distillate to a maximum of 50 percent volume fraction and the heavy distillate to a maximum of 50 percent.

2.4.5 *Premium diesel fuel*. This product is to be made from a blend of any of the six distillate materials, and must satisfy the following conditions: viscosity less than 2.6 centistokes at 100°F; specific gravity less than 0.840; cetane number greater than 55;

90 percent over in ASTM distillation at a maximum of 585°F; and end point at a maximum of 646°F. The gravity, viscosity and cetane number of any blends obey equations (VI.10), (VI.11), and (VI.12), respectively. By holding the light distillate to a minimum of 20 percent by volume, and the heavy distillate to a maximum of 33 percent, the specifications for the 90 percent and end point in ASTM distillation ought to be met.

2.4.6 *Number 6 bunker fuel oil.* This product is composed primarily of straight-run and cracked residues. The specifications are as follows: viscosity less than 375 centistokes at 122°F, and specific gravity less than 1.014. Since mixed residues, in general, do not meet these specifications, they are blended with distillate cutter stocks. As in the case of the gas oil products, the gravity and viscosity of the blends conform to equations (VI.10) and (VI.11).

2.5 CRACKED GAS PRODUCTS

The analysis of the cracked gases is taken to be the following:

Component	Weight fraction
Fuel gas (C2 and lighter)	0.285
Propane-propylene (C3 cut)	.218
Butane-butylene (C4 cut)	.497

The gases are separated into these three fractions — the light gas for fuel, the propane for sale, and the butane either for sale or for gasoline blending.

2.6 REVENUES AND PROCESSING COSTS

2.6.1 *Heat.* Since the cost of crude-topping remains constant, this element need not be taken into account here. The heat requirements for the cracking coils are estimated under the following assumptions. (A) It takes 32,000 British thermal units per pound mole of bonds broken. (B) The fraction vaporized in each stream is: light oil — 100 percent; medium — 85 percent; heavy — 60 percent; and residue — 40 percent. (C) There is no heat exchange between any refinery streams. If this approach is valid, the heat consumption works out as follows:

Total heat requirement (Btu per pound of crude)

$$= 38.1F_L + 31.9F_M + 52.3F_H + 55.9F_R$$
$$+ 286R_LZ_L + 340R_MZ_M + 431R_HZ_H \qquad \text{(VI.13)}$$

Heat costs $10\cancel{c}/10^6$ Btu.

2.6.2 *Steam.* Leaving out the value of the steam in the crude column and strippers, and considering only the steam in the cracked products column, the following development is used: 0.2 pound of steam per gallon for light, medium, and heavy distillates, and 0.4 pound of steam per gallon for the residue.

Steam requirement (pounds per pound of crude)

$$= 0.028 \, (Z_L + Z_M + Z_H) + 0.047C_R. \qquad \text{(VI.14)}$$

Steam costs $18\cancel{c}/10^3$ pounds. This steam is the exhaust from steam-driven pumps, and therefore the price includes pumping costs.

2.6.3 *Cooling water.* This water must condense steam, gasoline for reflux, the three cycle oil streams, and the residue.

Cooling water requirement (gallons per pound of crude)

$$= 0.455(0.146F_L + R_LZ_L) + 0.387(0.095F_M + R_MZ_M)$$
$$+ 0.267(0.123F_H + R_HZ_H) + 0.182(0.371F_R)$$
$$+ 0.107(Z_L + Z_M + Z_H) + 0.182C_R$$
$$+ 1.40C_{\text{Gaso}} \qquad \text{(VI.15)}$$

Recirculating cooling water costs $2\cancel{c}/10^3$ gallons.

2.6.4 *Lead requirements.* Ethyl fluid costs $0.25\cancel{c}$ per milliliter.

2.6.5 *Processing costs for propane and butane.* The cost of separating these cuts from their mixture with fuel gas, along with their subsequent liquefaction, is estimated at $0.7\cancel{c}$ per gallon of propane and butane.

2.6.6 *Definition of net revenues.* If the refiner were concerned with computing the net increase in product value over the raw materials and processing costs, it would be essential for him to take into account the value of the crude and the cost of primary fractionation. Since these elements are constant for all process variations considered here, it is necessary to consider only the gross value of the salable products over and above the processing costs listed in paragraphs 2.6.1–2.6.5.

2.7 EQUIPMENT CAPACITY RESTRICTIONS

The equipment that has been described is subject to specified maximum throughputs. The cracking coils, for example, have a certain effective area for heat transfer, and the furnaces are limited to a maximum heat load. Each of the three distillate cracking coils is designed so as to be capable of handling any amount up to 2.5 times the weight of the respective straight-run stream. Coil L, therefore, has a capacity of 2.5 (0.146) pounds of charge per pound of crude; coil M a capacity of 2.5 (0.095) pounds of charge; and coil H a capacity of 2.5 (0.123) pounds of charge.

The residuum cracking coil can process any amount up to the entire straight-run cut, 0.371 pound of residue per pound of crude.

The capacity of the fractionating column is a more complex matter than that of the coils. The most likely limit appears to be the "flooding" of the column at a certain vapor flow in the top section. This vapor is composed of gas, gasoline, and steam. The maximum vapor flow at this point is set at 85 percent of the flow with all four coils operating at maximum capacity, i.e. 98×10^{-4} moles per pound of crude. This brings about the following bound on the operable capacity of the secondary column:

Capacity of secondary column (10^{-4} moles per pound of crude)

$$
\begin{aligned}
= {} & 113(0.146F_L + R_LZ_L) + 97(0.095F_M + R_MZ_M) \\
& + 97(0.123F_H + R_HZ_H) + 3.4F_R \\
& + 16(Z_L + Z_M + Z_H) + 26C_R.
\end{aligned} \tag{VI.16}
$$

2.8 LIMITATIONS OF THE EQUATIONS

All through the development of this system of relationships, numerous simplifications had to be made. In a problem involving actual dollars and cents, it would be necessary to improve upon these simplifications. Instead of constant values for the yields and prices, alternative ones might have to be tried out. For instance, in the case of the yield structure from thermal cracking of the distillate streams, the numbers given in Table VI.3 are stated independently of the recycle ratio. This seems to be a fairly reasonable assumption for values of R_i between 0.5 and 0.9.

It is known, however, that as the ultimate conversion of a given stream goes up, the marginal output of gasoline goes down, and that of gas and residue increases. The yield estimates do not fully allow for this effect.

It is also worth recalling that the market values of the products were taken to be constant, regardless of the amounts sold. This approach may be valid enough if the output of the individual refinery is a small fraction of the production in a given region. If this is not true, the incremental value of the products would decline as additional quantities were placed upon the market.[2]

3. ECONOMIC AND TECHNOLOGICAL ASSUMPTIONS

Because of the fact that Singer spelled out the limitations of his analysis with great care, it is unnecessary to dwell at length upon them. Just as was done in the crude oil allocation, the naphtha reforming, and the gasoline blending problems described in previous chapters, he made the assumption that the prices realized upon the refined products would be independent of the quantities sold. He did not impose upper bounds upon the make of any product, nor did he require that the average refinery realizations decline with increases in the amounts that were sold.

From the economic engineering standpoint, perhaps the most serious defect of the model is the assumption of a standard percent "conversion-per-pass." In a more complete treatment of the subject, this conversion ratio might itself be determined within the framework of the calculation. Sachanen, citing W. L. Nelson, summarizes the pros and cons of increasing the crack-per-pass as follows:

1. Lower (ultimate) gasoline yield.
2. Higher octane number.
3. An increased tendency to produce coke.
4. The production of more gas.
5. Increased volatility of the light-end of the gasoline.[3]

---

[2] This point marks the end of Emanuel Singer's memorandum dated January 26, 1953, "Generalized Profitability Equations for a Simplified Refinery."

[3] A. N. Sachanen, *Conversion of Petroleum* (second edition, New York, 1948), p. 215.

This list of divergent effects is good evidence that there is a major economic balance problem involved in the choice of the optimum crack-per-pass. A fuller analysis of the cracking problem would include not only the recycle ratio, but also the temperature and duration of the cracking reaction as variables that are subject to control.

For a given recycle ratio and type of charge stock, the reaction time and temperature will determine the conversion results in noncatalytic, mixed-phase cracking.[4] If there were available any reliable data upon the interaction of these factors, the larger system would yield more useful results than the present one. Perhaps the development of improved computing techniques may stimulate refiners to gather additional information along these lines.

4. THE MATHEMATICAL MODEL

4.1 INTRODUCTION

The problem originally posed by Singer abounds with nonlinearities. The three simultaneous equations VI.1–VI.3 would in themselves be enough to require some computing technique along the lines of that which was used for the gasoline blending problem in Chapter V. Each of the remaining equations also contains nonlinear terms in the seven independent variables, F_i and R_i. The octane number-TEL blending relationships provide still further instances of a nonlinear structure. (See Figure 16.)

Fortunately, Kenneth Arrow was able to suggest a transformation of variables which — aside from the octane number curves — converted the problem into a standard linear programming form. Once discerned, the trick is obvious — to define seven new independent variables. The first four coincide with those that had been used previously, and each of the remaining three is formed by multiplying one of the old independent variables by a dependent one:

$$\begin{array}{lll} X_1 = F_L & X_5 = R_L Z_L & \text{(VI.17)} \\ X_2 = F_M & X_6 = R_M Z_M & \\ X_3 = F_H & X_7 = R_H Z_H. & \\ X_4 = F_R & & \end{array}$$

[4] Sachanen, pp. 182–225.

After these transformations have been carried through, each of the process equations from VI.1 through VI.16 becomes linear in the new independent variables. Under the convenient new definition, the independent variables represent *quantities* charged to the individual cracking coils, and not the recycle *ratios* generally employed by petroleum engineers. The change in definition does not alter any of the technical assumptions underlying the problem, but does make the optimizing calculation an easier one.

The only remaining nonlinearities are those associated with the ethyl blending chart. For purposes of the refinery-wide problem, it was decided to handle the TEL concentration level in the regular grade gasoline as an unknown parameter k_1, and to treat the other one as fixed at 2.80 ml per gallon.[5]

This latter TEL variable represents the only point at which a portion of the original problem was sacrificed in the interests of computing expediency. With this one exception, the entire network of relationships described by Singer may be represented by a system of the following standard form:

$$\text{Maximize} \quad \sum_{j=1}^{n} p_j x_j \qquad (j = 1, 2, \ldots, n) \qquad \text{(VI.18)}$$

$$\text{Subject to:} \quad \sum_{j=1}^{n} a_{ij} x_j = q_i \qquad (i = 1, 2, \ldots, m)$$

$$x_j \geq 0 \qquad (j = 1, 2, \ldots, n) \qquad \text{(VI.19)}$$

The group of conditions (VI.18) and (VI.19) is closely related to a system that has previously been discussed — that is, the setup for the crude oil allocation problem, equations (III.3). The earlier one was, in fact, a special case of this problem — an instance in which all the nonzero a_{ij} coefficients equal unity. In the preceding situation, the unique structure made it possible to compute the optimum program with great ease, but in the present case a more advanced computing algorithm is required — the "general simplex" method.

---

[5] Since the lead level in premium gasoline can only vary within the narrow limits of 2.80 and 3.00 milliliters, this did not seem important enough to warrant study as another unknown.

The simplex method is based upon three principles analogous to those discussed earlier: (1) Any schedule that satisfied the restrictions (VI.19) is termed a "feasible" one. If a feasible program contains *exactly* m nonnegative x_j activities, then it is called a "basic feasible" schedule. The set of m activities actually employed is termed the "basis."[6] (2) One of the optimal x_j schedules will always be a "basic feasible" solution. In case there are several alternative programs that are equally optimal — that is, each of which maximizes $\sum_j p_j x_j$ — at least one of these will be a basic solution. (3) For each basic feasible set of x_j, there may be defined a group of m independent dual variables, u_i. For each x_j activity *in* the basis, the dual variables are chosen in such a way that $\sum_i a_{ij} u_i = p_j$. Once an optimal basis is reached, these dual variables will exhibit a useful property. At such an optimum, for each x_j activity *outside* the basis, $\sum_i a_{ij} u_i \geq p_j$.

Just as in the crude oil allocation problem, there are three successive stages involved in determining the optimal set of x_j: first, finding a basic feasible solution; second, testing it for optimality through the u_i "shadow price" variables; and third, introducing into the basis any activity that displays a positive "surplus," i.e., one in which $\sum_i a_{ij} u_i < p_j$. The computation alternates between the second and third phase until an optimum is reached.

Since there is a voluminous literature on the simplex method of numerical solution, there is no need to deal with the subject here.[7] It suffices to note that George Dantzig and William Orchard-Hays have developed a technique that will handle any

[6] In order to constitute a "basis," one other technical condition must be satisfied. The matrix formed from the m activities must have a rank of m.

[7] The following represents a partial bibliography:

(1) *Activity Analysis of Production and Allocation*, ed. T. C. Koopmans (New York, 1951).

(2) Robert Dorfman, *Application of Linear Programming to the Theory of the Firm* (Berkeley, 1951).

(3) A. Charnes, W. W. Cooper, and A. Henderson, *An Introduction to Linear Programming* (New York, 1953).

system for $m \leq 40$ on a conventional IBM Card Programmed Calculator, Model 2.[8] With more advanced machines, larger structures may be analyzed.

In the case at hand, the Card Programmed Calculator's capacity turned out to be ample, for the problem may be reduced to a system with 26 equations and 61 x_j. Table VI.5 contains the complete group of p_j, a_{ij}, and q_i coefficients required for solving the optimum scheduling problem, equations (VI.18) and (VI.19). The x_j "activities" fall into three general classes:

1, 2, . . . , 7	Quantities of straight-run and recycle materials charged to individual cracking coils;
16, 21, . . . , 99	Components in various blended products;
101, 102, . . . , 126	Balance accounts.

The 26 equations in this instance also happen to be grouped under three headings:

1, . . . , 10	Availability and disposal of straight-run and cracked materials;
11, . . . , 14	Equipment capacity limitations;
15, . . . , 26	Specifications on various blended products.

In order to follow the transition from Singer's memorandum to the compact form of Table VI.5, the first step will be to define each of the independent variables, x_j, and then to show how these fit into the 26 restraint equations. Following this, the payoff coefficients, p_j, will be connected up with the process costs and the end-item revenues.

(4) John Chipman, "Computational Problems in Linear Programming," *The Review of Economics and Statistics*, vol. 35, no. 4 (November 1953).

(5) George B. Dantzig and William Orchard-Hays, "The Product Form for the Inverse in the Simplex Method," *Mathematical Tables and Other Aids to Computation*, vol. 8, no. 46 (April 1954).

[8] The number of potential activities, n, is not a limiting factor. For example, the Dantzig-Orchard-Hays algorithm could easily accommodate several hundred x_j — so long as the number of equations does not exceed 40.

Table VI.5. Coefficients for Programming a Simplified Thermal Cracking Refinery

Type of activity, x_j Index j	Quantities charged to individual cracking coils						
	F_L 1	F_M 2	F_H 3	F_R 4	$R_H Z_L$ (pounds) 5	$R_M Z_M$ (pounds) 6	$R_H Z_H$ (pounds) 7
Payoff coefficient, p_j	0.006 648	0.004 278	.005 456	.009 165	0.055 313	0.048 270	0.047 597
Index and purpose of restraint equation i							
Availability of materials (pounds):							
1. Straight-run gasoline	.146						
2. Straight-run light oil		.095					
3. Straight-run medium oil			.123				
4. Straight-run heavy oil							
5. Straight-run residue							
6. Cracked gasoline	−.0333	−.0191	−.0247	.371	−.1503	−.1440	−.1440
7. Cracked light oil	−.1022	−.00314	−.00332	−.0631	+.3000	−.0240	−.0240
8. Cracked medium oil	−.00175	−.0665	−.00185	−.0352	−.0390	+.3000	−.0360
9. Cracked heavy oil	−.00131	−.00114	−.0861	−.0390	−.0300	−.0420	+.3000
10. Cracked residue	−.0009	−.0009	−.0015	−.1850	−.0240	−.0420	−.0480
Equipment capacity limitations:							
11. Light distillate cracking coil (pounds)	.146				1.		
12. Medium distillate cracking coil (pounds)		.095				1.	
13. Heavy distillate cracking coil (pounds)			.123				1.
14. Secondary fractionating column (10^{-4} moles)	18.206	10.371	13.431	10.021	125.93	110.35	110.41
Product specifications:							
15. Regular gasoline (octane number)							
16. Number 1 fuel oil (viscosity-gallons)							
17. Number 2 fuel oil (viscosity-gallons)							
18. Number 2 fuel oil (cetane number-gallons)							
19. Number 2 fuel oil (flashpoint-gallons)							
20. Number 2 fuel oil (endpoint-gallons)							
21. Premium diesel fuel (viscosity-gallons)							
22. Premium diesel fuel (cetane number-gallons)							
23. Premium diesel fuel (specific gravity-gallons)							
24. Premium diesel fuel (90% point-gallons)							
25. Premium diesel fuel (endpoint-gallons)							
26. Number 6 fuel oil (viscosity-gallons)							

Table VI.5. Coefficients for Programming a Simplified Thermal Cracking Refinery (continued)

Type of activity, x_j Index j	Premium gasoline 16	Regular gasoline		Kerosene 52	Number 1 fuel oil			
		21	26		62	63	67	68
Payoff coefficient, p_j	11.0666	p_{21}	p_{26}	8.88	8.44	8.44	8.44	8.44
Index and purpose of restraint equation i								
Availability of materials (pounds):								
1. Straight-run gasoline		6.123						
2. Straight-run light oil				6.797	6.797			
3. Straight-run medium oil						6.914		
4. Straight-run heavy oil								
5. Straight-run residue								
6. Cracked gasoline	6.289		6.289					
7. Cracked light oil							6.997	
8. Cracked medium oil								7.205
9. Cracked heavy oil								
10. Cracked residue								
Equipment capacity limitations:								
11. Light distillate cracking coil (pounds)								
12. Medium distillate cracking coil (pounds)								
13. Heavy distillate cracking coil (pounds)								
14. Secondary fractionating column (10^{-4} moles)								
Product specifications:								
15. Regular gasoline (octane number)		1	$-k_2$					
16. Number 1 fuel oil (viscosity-gallons)					−0.022	0.185	−0.019	0.180
17. Number 2 fuel oil (viscosity-gallons)								
18. Number 2 fuel oil (cetane number-gallons)								
19. Number 2 fuel oil (flashpoint-gallons)								
20. Number 2 fuel oil (endpoint-gallons)								
21. Premium diesel fuel (viscosity-gallons)								
22. Premium diesel fuel (cetane number-gallons)								
23. Premium diesel fuel (specific gravity-gallons)								
24. Premium diesel fuel (90% point-gallons)								
25. Premium diesel fuel (endpoint-gallons)								
26. Number 6 fuel oil (viscosity-gallons)								

Table VI.5. Coefficients for Programming a Simplified Thermal Cracking Refinery (continued)

| | Components in various blended products (gallons) | | | | | | | | | | | |
| | Number 2 fuel oil | | | | | | Premium diesel fuel oil | | | | | |
Type of activity, x_j — Index j	72	73	74	77	78	79	82	83	84	87	88	89
Payoff coefficients, p_i	7.88	7.88	7.88	7.88	7.88	7.88	8.63	8.63	8.63	8.63	8.63	8.63
Index and purpose of restraint equation i												
Availability of materials (pounds)												
1. Straight-run gasoline	6.797						6.797					
2. Straight-run light oil		6.914						6.914				
3. Straight-run medium oil			7.122						7.122			
4. Straight-run heavy oil												
5. Straight-run residue												
6. Cracked gasoline												
7. Cracked light oil				6.997						6.997		
8. Cracked medium oil					7.205						7.205	
9. Cracked heavy oil						7.630						7.630
10. Cracked residue												
Equipment capacity limitations:												
11. Light distillate cracking coil (pounds)												
12. Medium distillate cracking coil (pounds)												
13. Heavy distillate cracking coil (pounds)												
14. Secondary fractionating column (10^{-4} moles)												
Product specifications:												
15. Regular gasoline (octane number)												
16. Number 1 fuel oil (viscosity-gallons)												
17. Number 2 fuel oil (viscosity-gallons)	−0.262	−0.055	0.116	−0.259	−0.060	0.190						
18. Number 2 fuel oil (cetane number-gallons)	−13	−18	−21	5	−5	0						
19. Number 2 fuel oil (flashpoint-gallons)	1	−1	−1	−1	−1	−1						
20. Number 2 fuel oil (endpoint-gallons)	1	−1	−1	−1	−1	1						
21. Premium diesel fuel (viscosity-gallons)							−0.132	0.075	0.246	−0.129	0.070	0.320
22. Premium diesel fuel (cetane number-gallons)							−2	−3	−6	20	20	15
23. Premium diesel fuel (specific gravity-gallons)							−0.024	−0.010	0.015	0	0.025	0.076
24. Premium diesel fuel (90% point-gallons)							−.8	.2	.2	−0.8	.2	.2
25. Premium diesel fuel (endpoint-gallons)							−.33	−.33	.67	−.33	−.33	.67
26. Number 6 fuel oil (viscosity-gallons)												

131

Table VI.5. Coefficients for Programming a Simplified Thermal Cracking Refinery (continued)

| | Components in various blended products (gallons) | | | | | | | |
| | Number 6 fuel oil | | | | | | | |
Type of activity, x_j Index j	90	92	93	94	95	97	98	99
Payoff coefficient, p_j	2.26	2.26	2.26	2.26	2.26	2.26	2.26	2.26
Index and purpose of restraint equation i								
Availability of materials (pounds):								
1. Straight-run gasoline								
2. Straight-run light oil		6.797	6.914					
3. Straight-run medium oil				7.122				
4. Straight-run heavy oil					7.864			
5. Straight-run residue								
6. Cracked gasoline								
7. Cracked light oil						6.997		
8. Cracked medium oil							7.205	
9. Cracked heavy oil								7.630
10. Cracked residue	8.513							
Equipment capacity limitations:								
11. Light distillate cracking coil (pounds)								
12. Medium distillate cracking coil (pounds)								
13. Heavy distillate cracking coil (pounds)								
14. Secondary fractionating column (10^{-4} moles)								
Product specifications:								
15. Regular gasoline (octane number)								
16. Number 1 fuel oil (viscosity-gallons)								
17. Number 2 fuel oil (viscosity-gallons)								
18. Number 2 fuel oil (cetane number-gallons)								
19. Number 2 fuel oil (flashpoint-gallons)								
20. Number 2 fuel oil (endpoint-gallons)								
21. Premium diesel fuel (viscosity-gallons)								
22. Premium diesel fuel (cetane number-gallons)								
23. Premium diesel fuel (specific gravity-gallons)								
24. Premium diesel fuel (90% point-gallons)								
25. Premium diesel fuel (endpoint-gallons)								
26. Number 6 fuel oil (viscosity-gallons)	0.1354	−0.9324	−0.7075	−0.5325	−0.1215	−0.9029	−0.6963	−0.4527

132

Table VI.5. Coefficients for Programming a Simplified Thermal Cracking Refinery (continued)

Type of activity, x_j / Index j	101	102	103	104	105	106	107	108	109	110	111	112	113	114	115	116	117	118	119	120	121	122	123	124	125	126	Right-hand side q_i
Payoff coefficient, p_j																											
Index and purpose of restraint equation i																											
Availability of materials (pounds):																											
1. Straight-run gasoline	1																										0.248
2. Straight-run light oil		1																									.146
3. Straight-run medium oil			1																								.095
4. Straight-run heavy oil				1																							.123
5. Straight-run residue					1																						.371
6. Cracked gasoline						1																					0
7. Cracked light oil							1																				0
8. Cracked medium oil								1																			0
9. Cracked heavy oil									1																		0
10. Cracked residue										1																	0
Equipment capacity limitations:																											
11. Light distillate cracking coil (pounds)											1																.3650
12. Medium distillate cracking coil (pounds)												1															.2375
13. Heavy distillate cracking coil (pounds)													1														.3075
14. Secondary fractionating column (10^{-4} moles)														1													98
Product specifications:																											
15. Regular gasoline (octane number)															1												0
16. Number 1 fuel oil (viscosity-gallons)																1											0
17. Number 2 fuel oil (viscosity-gallons)																	1										0
18. Number 2 fuel oil (cetane number-gallons)																		1									0
19. Number 2 fuel oil (flashpoint-gallons)																			1								0
20. Number 2 fuel oil (endpoint-gallons)																				1							0
21. Premium diesel fuel (viscosity-gallons)																					1						0
22. Premium diesel fuel (cetane number-gallons)																						1					0
23. Premium diesel fuel (specific gravity-gallons)																							1				**0**
24. Premium diesel fuel (90% point-gallons)																								1			**0**
25. Premium diesel fuel (endpoint-gallons)																									1		0
26. Number 6 fuel oil (viscosity-gallons)																										1	0

4.2 DEFINITION OF THE x_j "ACTIVITIES"

4.2.1 *Cracking coil inputs.* Equations (VI.17) have already defined the first seven x_j, the quantities of material charged to individual cracking units. Once these seven levels are determined, it is possible to calculate the amounts of each intermediate material available for blending into the end products. For example, equation (VI.8) indicates that if F_L is set equal to unity (that is, if the entire 0.146 pounds of the straight-run light distillate fraction are cracked), there is a yield of 0.0333 pounds of cracked gasoline per pound of crude input to the refinery. This coefficient is entered in column 1, line 6, of Table VI.5, and the yields of other cracked products from this same process are entered on lines 7, 8, 9, and 10.[9] (See equations VI.1 through VI.6 and VI.9 above.)

According to equations (VI.4) through (VI.6), in order to calculate the *net* effect of recycling the three cracked distillate oils, it is necessary to deduct the cycle oil yield of 70 percent per pass from the gross charge of 100 percent. In other words, each of the three recycling activities — x_5, x_6, and x_7 — on balance consume 30 percent of the gross charge. The constants a_{75}, a_{86}, and a_{97} all equal $+0.3000$, in accordance with the convention upon the sign of input coefficients.

4.2.2 *Components of blends.* In order to be converted into end items, the various intermediate oils produced by primary distillation and by the cracking processes must be blended together. For this stage of the analysis, it is necessary to consider eight of the ten possible end products listed in the beginning of this chapter. (The propane and the butane gases are salable directly, and do not enter here.) The eight marketable oils must be made up from the five straight-run and the five cracked materials listed in rows 1–10 of Table VI.5.

Each possible component is given a two-digit subscript. The first digit indicates the end item, and the second digit the intermediate material used for blending.[10] The subscript 63, for ex-

[9] By convention here, the net output of an intermediate product is shown with a negative sign, and the inputs with positive coefficients. This seemingly unnatural convention arises out of the fact that the balance accounts must not be negative.

[10] In order to avoid a three-digit subscript, the component x_{90} is to be interpreted as the quantity of stock 10 (cracked residue) employed in product 9 (Number 6 fuel oil).

ample, indicates that the particular component consists of oil 3 (medium straight-run), and that it is to be blended into end product 6 (Number 1 fuel oil).

Two of the end products — kerosene and premium gasoline — contain only one possible intermediate. Number 6 fuel oil, on the other hand, could conceivably be made up from any combination of the eight nongasoline materials.[11]

Since the end products are sold in terms of volumetric units, each of the blending components is measured in terms of "gallons of product per pound of crude oil." With this convention, it is necessary to translate between pounds of intermediate oils and gallons of end products. The cracked gasoline, for example, has a specific gravity of 0.755, and there are consequently 6.289 pounds of gasoline per gallon. This volumetric conversion factor determines the input coefficients on line 6 for the two cracked-gasoline-using activities — numbers 16 and 26. All of the other end-item components enter the material balance equations in a similar way, i.e., through the weight-to-volume conversion coefficients.

4.2.3 *Balance accounts.* Each of the variables numbered from 101 through 126 represents a "disposal activity" for one of the 26 equations. The unknown x_{101}, for example, represents the excess (if any) of light straight-run gasoline over the amount actually employed in the regular grade product. Similarly, x_{106} represents the net surplus of cracked gasoline production over the amounts blended into finished products, and the unknown x_{111} the surplus of light distillate cracking capacity.

In many cases, these balance activities turn out to be operated at a zero level in an optimal solution. They do constitute an important element of refinery flexibility, however, and should not be discarded summarily. In addition, the existence of this 26-order identity matrix makes it possible to cut down on the computational requirements. By starting off from an initial basis of activities 101, 102, . . . , 126, it is possible to eliminate the tedious job of determining a *basic* feasible solution.

Just as in the crude oil allocation problem, Chapter III, it is

[11] By contrast with the Union Oil Company blending problem, there are no gasoline cutter stocks employed here. The gasolines are not distilled in such a way that the 300°–400° heavy material is available for cutter purposes.

assumed that these "give-away" activities entail a zero net penalty to the refiner. In some cases, to be realistic, it would be necessary to associate a penalty with them. If there were a surplus of cracked residue, the refiner could not pump it into the sea. He might have to incur some expense in order to consume the oil as waste material. This particular surplus, incidentally, arose in connection with only one of the four calculations presented below, in run number 3. An excess of cracked residue did not occur in any of the cases where a conventional price structure was employed.

4.3 DEFINITION OF THE 26 EQUATIONS

4.3.1 *Availability of materials.* Each of the first ten rows in Table VI.5 is connected with the production and the consumption of one particular intermediate oil. An engineer will generally refer to these conditions as "material balance" equations. In all ten rows, the quantities are measured in "pounds of intermediate oil per pound of crude oil charged to the refinery."

Line 2, for example, is associated with the straight-run light distillate. On the last page of Table VI.5, under the column headed "Right-hand side, q_i," one finds that there are 0.146 pounds of this material available per pound of crude oil. The distillate may be employed either as cracking stock (column 1), as kerosene (column 52), or as a component in one of four fuel oils (columns 62, 72, 82, or 92). The sum of these various uses of light distillate oil, plus the excess (column 102), must always equal the amount available from crude distillation — 0.146 pounds.

Since the cracked materials are, by definition, produced in the cracking coils, and not by primary crude distillation, there is no "outside" input of these items. The right-hand entry, q_i, of rows 6–10 equals zero in each case.

Inspecting line 7, the reader will see that cracked light distillate oil may originate in any one of six cracking activities — columns 1–4 and columns 6 and 7. It is consumed in one of the recycling operations (column 5), and is a possible component of the four fuel oils (columns 67, 77, 87, and 97). The remainder enters into the balance account (column 107). Again with this cracked distillate, there is a material balance condition. The amounts produced must equal the amounts consumed plus the (nonnegative) balance.

4.3.2 *Equipment capacity limitations.* In setting up an optimum refinery operations schedule, it is necessary to take account not only of the availability of raw materials, but also of the capacity of the existing equipment. From section 2.7 above, it will be recalled that the three distillate cracking coils were each designed so as to handle any gross throughput up to 2.5 times the weight of the individual straight-run material. These given capacities are entered on the right-hand side of lines 11–13. The coefficients in each of the three rows refer to the weight of the gross throughput. If, say, the entire light virgin distillate fraction is cracked (i.e., if x_1 turns out to be unity), this will take up 0.146 pounds of cracking charge capacity. Accordingly, the coefficient in column 1, row 11 is set at 0.146. The corresponding recycle operation (column 5) has already been stated in terms of the weight of gross charge, and for this reason its coefficient in line 11 becomes unity.

Row 14, dealing with the capacity of the secondary fractionating column, is derived from Emanuel Singer's equation (VI.16). In order to obtain the one set of coefficients from the other, it is necessary to substitute the explicit definitions of Z_L, Z_M, Z_H, and C_R given previously in equations (VI.1) through (VI.3) and (VI.9). Each of these four dependent variables may be determined through a linear combination of the seven unknowns — $x_1, x_2, \ldots, x_7$.

The process of algebraic substitution is a long-winded one, but is elementary and will not be carried out within these pages. Let it suffice to say that the coefficients in line 14 are the result of this straightforward elimination, and that they measure the total use of secondary fractionating capacity associated with each of the seven cracking operations. These various uses — together with the idle capacity (column 114) — must equal the prescribed level of 98 10^{-4} moles of vapor flow per pound of crude oil charge. The constant of 98 is entered on the right-hand side as q_{114}.

It may be wondered why there is no capacity restriction formally associated here with the residue cracking coil. But according to section 2.7 above, this unit was initially designed so as to be capable of processing any amount up to the maximum charge stock available — 0.371 pounds of straight-run material. The residuum cracking coil cannot be a limiting factor in the static problem that is studied initially. If the thermal cracking equipment were being newly built or if a heavier type of crude oil were being processed, the capacity of this coil would have to be treated as an additional bottleneck.

4.3.3 *Product specifications.* Equations VI.15 through VI.26 serve to ensure that each of the blended end products will meet the specifications laid down in section 2.4. In a number of cases, it turned out to be possible to avoid setting in these requirements as a row in Table VI.5. With premium grade gasoline, for example, the 2.8 ml per gallon TEL concentration level will suffice to meet the 88 octane number requirement. Accordingly, the premium gasoline product is debited with the cost of 2.8 ml of TEL, and the knock-rating specification is met exactly.

Regular grade gasoline. The octane number of this product is controlled through row 15. The unknown variable k_1 represents the ethyl fluid concentration in regular grade gasoline. This unknown controls the coefficient k_2, the maximum ratio of straight-run to cracked gasoline that will ensure the blend meeting the regular gasoline octane number requirement of 82. The ratio k_2 may be determined through inspection of Figure 16 for various TEL concentration levels, k_1. The ethyl fluid concentration not only affects k_2, but also p_{21} and p_{26}, the net refinery realization on regular grade gasoline. Table VI.6 lists these unknown parameters for five different values of k_1.

After the entire numerical analysis had been performed upon this linear programming problem, a less awkward method was discovered for handling the TEL concentration levels. Instead of representing an activity of *blending* component X into product Y, the columns could equally well refer to alternative recipes for producing one gallon of product Y. In each case, the recipe would call for a different TEL level. For example, at a 3.0 ml per gallon concentration, the regular gasoline blend would consist of 57 percent straight-run plus 33 percent cracked gasoline plus 10 percent butane. For the 2.0 ml level activity, the batch would contain 39 percent of straight-run, 51 percent of cracked gasoline, and 10 percent of butane.

Table VI.6. Five Ethyl Fluid Concentration Levels and Corresponding Values for k_2, p_{21}, and p_{26}

k_1 (milliliters of TEL per gallon of 82 octane number gasoline)	Maximum percent straight-run in blended 82 octane number gasoline (Figure 16)	Minimum percent cracked gasoline in blended 82 octane number gasoline	k_2 (maximum ratio of straight-run to cracked component)	p_{21} and p_{26} (net refinery realization on regular grade gasoline, after butane credit and TEL debit)
3.0	57.0	33.0	1.727	9.7666
2.9	55.5	34.5	1.609	9.7944
2.5	50.0	40.0	1.250	9.9055
2.0	39.0	51.0	0.7647	10.0444
1.0	10.0	80.0	0.1250	10.3222

Through pre-solving part of the problem in this manner, it is possible to avoid calculating separate schedules for each TEL concentration case. This approach eliminates any need for row 15, dealing with the octane number of regular grade gasoline. And in addition, it would be possible to consider a 3.0 ml concentration level for the premium product. In retrospect, it seems unfortunate that the problem was not handled through pre-selecting the *set* of acceptable blends. This device turns out to be quite useful whenever (a) there is a small number of possible components for the blend, or (b) all but one or two of the possible components lie on the same side of a specification limit.

Number 1 distillate fuel oil. Line 16 serves to ensure that this product will meet a maximum viscosity of 1.9 centistokes at 100°F. The coefficients in columns 62, 63, 67, and 68 represent $(\mu_i - \mu_{spec})$ — the algebraic excess of the viscosity blending number of the particular component i over the required viscosity blending number for Number 1 fuel oil. These four constants, together with the nonnegative balance account (column 116), ensure that the viscosity of the blend will not exceed the specified upper limit. The coefficients $(\mu_i - \mu_{\text{spec}})$ are entirely analogous to the d_{ji} and e_{ji} constants associated, respectively, with the sulfur and the volatility specifications in the Union Oil Company's gasoline blending problem.

The Number 1 fuel oil end product must not only meet the viscosity requirement, but also a maximum specific gravity of 0.850. Since the viscosity specification cannot be satisfied without *at the same time* meeting the gravity specification, this latter condition does not enter into the problem as a formal requirement. The proof of this assertion is to be found in Figure 17, which plots the viscosity blending number, μ_i, versus the specific gravity for the four possible components of Number 1 fuel oil. The irregular shaded polygon with vertices 2-3-8-7 contains the specific gravities and the viscosity blending numbers for all conceivable blends of these four components. The quadrant labelled "I" contains all points that meet neither the gravity nor the viscosity requirement of the end product. The points in quadrant "II" satisfy the gravity, but not the viscosity specification; those in quadrant III meet both; and those in quadrant IV meet the viscosity, but not the gravity. No points of the shaded polygon are to be found in quadrant IV. It follows that no blend will satisfy the viscosity requirement without also falling within the prescribed gravity limit. This side-calculation makes it possible to avoid inserting an extra row in Table VI.5, and to that extent reduces the burden of the machine computations.

Number 2 distillate fuel oil. The viscosity of this item is controlled through line 17. The coefficients again represent $(\mu_i - \mu_{\text{spec}})$, and together with the nonnegative balance account ensure that the blend of six components (72, 73, 74, 77, 78, and 79) will not exceed the limit of 4.3 centistokes at 100°F. A two-dimensional comparison similar to that of Figure 17 reveals that this viscosity specification will also bring the blend within the required specific gravity limit of 0.882.

The cetane number is handled through equation 18. Here, since the cetane requirement is a lower, rather than an upper limit, the coefficients represent the algebraic excess of the specification cetane number over that of the particular component i: $(\text{CN}_{\text{spec}} - \text{CN}_i)$.

The flash point is controlled by holding the light distillate to a

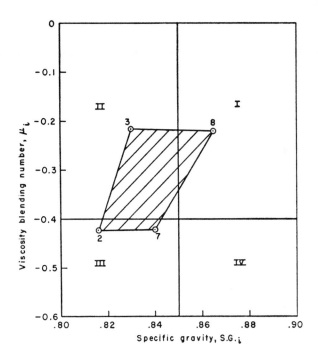

Fig. 17. Properties of Component Materials for Number 1 Fuel Oil

maximum of 50 percent volumetric fraction. This condition, maintained by row 19, is the same as the following:

$$x_{72} + x_{77} \leq x_{73} + x_{74} + x_{78} + x_{79}. \qquad \text{(VI.20)}$$

In a similar manner, line 20 takes care of the end-point requirement — that the heavy distillate be held to a maximum of 50 percent by volume. Or,

$$x_{74} + x_{79} \leq x_{72} + x_{73} + x_{77} + x_{78}. \qquad \text{(VI.21)}$$

Premium diesel fuel. Row 21 guards against an off-specification viscosity for this product. The coefficients again represent $(\mu_i - \mu_{\text{spec}})$. The cetane number requirement is maintained through line 22, where the coefficients are determined by $(CN_{\text{spec}} - CN_i)$.

In the case of the premium diesel product, unfortunately the two-dimensional plots of gravity versus viscosity, and of gravity versus cetane number, do not rule out the possibility that the specific gravity may become an effective limitation upon the operating schedule. For this reason, it is necessary to devote one row in the matrix VI.5 to this particular purpose. The gravity requirement represents an upper limit, and the coefficients in line 23 are defined by: $(S.G._i - S.G._{\text{spec}})$.

The 90 percent point in ASTM distillation is to be met by maintaining a minimum of 20 percent of light oils in the blend. Row 24 is equivalent to the following:

$$x_{82} + x_{83} \geq 0.2(x_{82} + x_{83} + x_{84} + x_{87} + x_{88} + x_{89}). \qquad \text{(VI.22)}$$

The ASTM distillation end-point requirement is to be satisfied by holding the heavy distillate to a maximum of 33 percent. Line 25 of the matrix is derived through transposing terms in the following inequality:

$$x_{84} + x_{89} \leq 0.33(x_{82} + x_{83} + x_{84} + x_{87} + x_{88} + x_{89}). \qquad \text{(VI.23)}$$

Number 6 bunker fuel oil. Here, the viscosity requirement is a maximum of 375 centistokes at 122°F. The last line of the matrix contains the coefficients $(\mu_i - \mu_{\text{spec}})$ that check out this property. The gravity specification on Number 6 fuel oil is excluded for the same reason as in the case of the Number 1 and Number 2 oils. It is impossible for a blend of these particular components to meet the viscosity specification without also satisfying the gravity limit.

4.4 DEFINITION OF THE p_j, PAYOFF COEFFICIENTS

4.4.1 $p_1, p_2, \ldots, p_7$. The payoff coefficients that go along with columns 1–7 represent the excess of cracked gas revenues over the utility expenses per unit of the particular activity. From equation (VI.7), we already know how to calculate the output of cracked gases as a linear function of each of the seven basic operations. Section 2.5 has indicated the composition of this stream; section 2.6.5 gives the processing costs associated with propane and butane; and Table VI.4 shows the market price of propane, butane, and fuel gases. When combined together, these pieces of information reveal that the gas stream yields the refiner a net value of 1.1144¢ per pound of cracked gases. Upon applying this figure to the coefficients of equation (VI.7), one obtains line 1 of Table VI.7 — the gas revenues net of gas processing costs.

From section 2.6.1, it is possible to calculate the heat requirements as a linear function of the seven individual cracking operations. When priced out at 10¢ per million Btu, the second line of Table VI.7 is determined — the process heat expenses associated with cracking each oil.

In order to estimate the steam requirements, it is necessary to substitute the explicit definitions of Z_L, Z_M, Z_H, and C_R in equation (VI.14). Once this substitution is performed, it is again possible to state the steam requirements as a linear function of the seven independent variables, $x_1, x_2, \ldots, x_7$. If the steam is priced at 18¢ per thousand pounds, these costs will be as indicated by the third line of Table VI.7.

The cooling water requirements may be determined by a similar procedure of substituting explicit terms in equation (VI.15) in place of the dependent variables Z_L, Z_M, Z_H, C_R, and C_{Gaso}. Charging up the cooling water at 2¢ per thousand gallons, this calculation gives the fourth line of Table VI.7.

The various utility expenses are now subtracted off from the net gas revenues to yield line 5 — the p_j coefficients associated with columns 1–7. It is worth noting that the gas revenues exceed the combined utility expenses by a factor of approximately ten.

Table VI.7. Calculation of $p_1, p_2, \ldots, p_7$ Coefficients

Charge stock Unknown quantity of charge (cents per unit of charge)	SR, L x_1	SR, M x_2	SR, H x_3	SR, R x_4	Cr, L x_5	Cr, M x_6	Cr, H x_7
1. Gas revenues, net of gas processing costs	0.007 322	0.004 770	0.006 173	0.010 341	0.060 179	0.053 491	0.053 491
2. Heat expenses	.000 381	.000 319	.000 523	.000 559	.002 860	.003 400	.004 310
3. Steam expenses	.000 056	.000 038	.000 049	.000 214	.000 382	.000 401	.000 403
4. Cooling water expenses	.000 237	.000 135	.000 145	.000 403	.001 624	.001 420	.001 181
5. Gas revenues net of all processing costs, p_i	.006 648	.004 278	.005 456	.009 165	.055 313	.048 270	.047 597

Furthermore, the heat costs generally exceed 50 percent of the total utility charges.

4.4.2 p_{16}, p_{21}, and p_{26}. In calculating the net value to the refiner of one gallon of the cracked component in premium grade gasoline, it is necessary to take account of the fact that this material is to be mixed with butane and with ethyl fluid. The value of the end product, premium grade gasoline, is 11.56¢ per gallon. If it is to be blended with 2.8 milliliters of ethyl fluid, the premium gasoline becomes worth 10.86¢ per gallon net of lead costs. For each gallon of the thermally cracked component, the refiner can promote 0.1111 gallons of 9¢ butane into premium grade gasoline. The final netback on the cracked component comes out:

$$p_{16} = 1.1111 \left[11.56 - 2.8 \, (0.25)\right] - 0.1111 \, (9)$$
$$= 11.0666 ¢ \text{ per gallon.} \quad \text{(VI.24)}$$

In the case of the regular grade gasoline, the refiner is again able to convert $\frac{1}{9}$ of a gallon of butane into motor fuel for each gallon of straight-run or thermally cracked gasoline component. With this end product, however, the lead concentration level, k_1, is an unknown variable that is to be determined in the course of the optimization. Taking account of the butane profit and of the TEL costs, the coefficients p_{21} and p_{26} become linear functions of the TEL concentration level:

$$p_{21} = p_{26} = 1.1111 \left[10.44 - k_1 \, (0.25)\right] - 0.1111 \, (9)$$
$$= 10.6 - 0.2778 \, k_1 ¢ \text{ per gallon.} \quad \text{(VI.25)}$$

Equation (VI.25), solved for five alternative TEL levels, yields the right-hand column of Table VI.6.

4.4.3 p_{52}, p_{62}, p_{63}, . . . , p_{126}. The p_j coefficients in columns 52, 62, 63, . . . , 99 measure the refinery gate payoff of products 5 through 9 — that is, of kerosene, Number 1 oil, Number 2 oil, premium diesel, and Number 6 bunker oil. The prices are each stated in cents per gallon, and are taken over directly from Table VI.4.

The payoff constants in columns 101, 102, . . . , 126 — the balance accounts — are all set at zero. In other words, neither a cash penalty nor a gain is ascribed to any of the disposal activities.

4.5 A SUMMARY OF THE MATHEMATICAL MODEL

During the course of the preceding pages, it has been shown how the translation was effected from Singer's original engineering problem into the standard linear programming form. Each of the parts — the A matrix, the right-hand side Q vector, and the payoff vector P — was derived from the conditions stated by him. It now remains to solve for the unknowns, the X vector and the k_1 parameter, that will maximize the payoff subject to the stated constraints. In addition, the parametric linear programming technique will be employed in order to determine the effects of varying certain of the initial assumptions.

5. THE NUMERICAL ANALYSIS, RUN NUMBER 1

5.1 THE BASIC CASE, PRODUCT-MIX SUMMARY

Using the set of coefficients indicated in Table VI.5, the original linear programming problem was solved on an IBM Card Programmed Calculator. In place of the conventional simplex method, the Dantzig-Orchard-Hays modification was employed, carrying the product form of the inverse. Forty-three iterations were required in order to obtain an optimum refinery schedule, and the running time involved was approximately ten machine hours.

The product-mix resulting from this computation is summarized in Table VI.8. For comparative purposes, these numbers are listed alongside the corresponding ones for the actual output experienced in the vicinity of the hypothetical refinery — in the Oklahoma, Kansas, and Missouri region during 1952–53. The optimal make of the main product, butanized motor gasoline, checks closely with that given for the entire region — a total of 54.10 percent versus 57.15 percent by volume of the crude. Assuming that the Oklahoma refiners were actually maximizing their immediate profits during this period, the gasoline forecast is remarkably close.

Unfortunately for predictive purposes, the production of distillate fuels does not follow the same item-by-item pattern as that for the region as a whole. Neither kerosene nor Number 1 fuel

Table VI.8. Product-Mix Summary, Run No. 1

Net output (percent by volume of crude distilled)	Run number 1 (basic case)	Oklahoma, Kansas, Missouri region July, 1952 through June, 1953[a]
Butanized gasoline:		
Premium	4.69	—
Regular	49.41	—
Total gasoline	54.10	57.15
Distillate oils:		
Kerosene	0	2.54
Number 1 fuel oil	0	⎫
Number 2 fuel oil	9.00	⎬ 23.55
Premium diesel	17.40	⎭
Total distillates	26.40	26.09
Residual fuel oil (Number 6 grade)	19.80	9.71

[a] U. S. Bureau of Mines, *Monthly Petroleum Statements 355–366* (July 1952 through June 1953).

oil is scheduled at all, and there is a comparatively large output of premium diesel fuel. Interestingly enough, the *total* output of these oils, 26.40 percent of the crude, almost coincides with the total of 26.09 percent for the entire region. This result could be a sheer coincidence, but it may also reflect the fact that the various grades of distillate materials are close substitutes as far as the refiner is concerned.

The outstanding discrepancy between the optimal program of run number 1 and that which was actually recorded for the Oklahoma district showed up in connection with the Number 6 residual fuel oil product. The basic case calls for over twice as much of this end item as was in fact produced during the 1952–53 period. This result is probably linked to the fact that the thermal cracking refinery contains no equipment for coking, visbreaking, or vacuum distillation, but might also arise from any of the other factors not included in the present analysis. In order to track down the reasons for the two-to-one discrepancy, further research would be needed on the marketing problems as well as on the technology considered here.

5.2 RUN NUMBER 1, DETAILS OF THE OPTIMAL PROGRAM

Table VI.9 provides the "activity levels" and the "shadow prices" that go along with the solution to the initial linear programming problem. Three TEL levels for the regular grade gasoline were considered: $k_1 = 1.0$ ml per gallon, 2.9 ml, and 3.0 ml. Once the first case had been solved, it was no longer necessary to repeat the entire linear programming iterative routine. Instead, an elimination equation made it possible to introduce the new x_{26} vector (corresponding to the altered value of k_1) into the basis, to drive out the old vector, and to test for optimality.

As might have been anticipated from the particular configuration of product prices and gasoline qualities, the most profitable of the three cases was the 3.0 ml concentration. The greater the TEL level here, the higher the payoff of the refinery program. Remarkably enough, in this instance the gasoline blending problem seems to be divorced entirely from the other operations of the refinery. The only difference between the three cases occurs in the gasoline-consuming activities — 16, 21, 26, and 101. The other refining processes — the cracking, the recycling, and the fuel oil blending — are completely unaffected by the ethyl blending level. For this reason, during the subsequent computing runs, numbers 2 through 4, the parameter k_1 was held constant at 3.0 ml per gallon, and no attempt was made to include it as an independent variable.

Table VI.9, when inspected with care, yields a number of insights into the refining problem. Of the first four x_j — the cracking of straight-run oils — there is only one instance in which the optimum activity level turns out to be unity. That is, it is only in the case of the residual oil cut that it appears desirable to charge the entire material into the cracking coil. The other three straight-run stocks all have profitable uses in fuel oil blending as well as in cracking.

Activities 5, 6, and 7 are all concerned with the cycling of cracked streams back into the conversion units. Apparently, under the assumptions set into this problem, it pays to recycle

Table VI.9. Solution to Run No. 1, Basic Case, $k_1 = 3.0$ ml TEL
per Gallon of Regular Grade Gasoline

j	"Activity levels," x_j		i	"Shadow prices," u_i	
1	0.35546	F_L	1	1.4722	(¢/pound, SR gasoline)
2	0.80403	F_M	2	1.3643	(¢/pound, SR light oil)
3	0.32506	F_H	3	1.1987	(¢/pound, SR medium oil)
4	1.0000	F_R	4	1.0475	(¢/pound, SR heavy oil)
5	0.17063	$R_L Z_L$ (pounds)	5	0.73807	(¢/pound, SR residue)
6	.16112	$R_M Z_M$ (pounds)			
7	.26752	$R_H Z_H$ (pounds)	6	1.7597	(¢/pound, cracked gasoline)
			7	1.2906	(¢/pound, cracked light oil)
16	.0060736	(gal of stock 6 in product 1)	8	1.1238	(¢/pound, cracked medium oil)
			9	0.97556	(¢/pound, cracked heavy oil)
21	.040503	(gal of stock 1 in product 2)	10	0.14621	(¢/pound, cracked residue)
26	.023453	(gal of stock 6 in product 2)			
			11	0	(¢/pound capacity, coil L)
68	0	(gal of stock 8 in product 6)	12	0.034527	(¢/pound capacity, coil M)
			13	0.078693	(¢/pound capacity, coil H)
74	.0044252	(gal of stock 4 in product 7)	14	0.000073173	(¢/10^{-4} moles capacity,
78	.0085554	(gal of stock 8 in product 7)			secondary column)
82	.013845	(gal of stock 2 in product 8)	15	0.75275	(¢/gal, regular gasoline)
83	.0026927	(gal of stock 3 in product 8)	16	1.9042	(¢/μ-gal, Number 1 fuel oil)
84	.0072314	(gal of stock 4 in product 8)	17	3.6209	(¢/μ-gal, Number 2 fuel oil)
87	.0011888	(gal of stock 7 in product 8)	18	0	(¢/CN-gal, Number 2 fuel oil
			19	0	(¢/gal, Number 2 fuel oil)
90	.024696	(gal of stock 10 in product 9)	20	0	(¢/gal, Number 2 fuel oil)
97	.0037034	(gal of stock 7 in product 9)	21	5.0646	(¢/μ-gal, diesel fuel oil)
			22	0.012647	(¢/CN-gal, diesel fuel oil)
111	.14248	(pounds capacity, coil L)	23	0	(¢/S.G.-gal, diesel fuel oil)
118	.050152	(CN-gal of Number 2 fuel oil)	24	0	(¢/gal, diesel fuel oil)
119	.012981	(flashpoint-gal of Number 2	25	0	(¢/gal, diesel fuel oil)
		fuel oil)	26	7.4985	(¢/μ-gal, Number 6 fuel oil)
120	.0041302	(endpoint-gal of Number 2 fuel			
		oil)		(payoff) $= \sum_i p_i x_i = 1.1204$	(¢/pound of crude
123	.00025073	(S.G.-gal of diesel fuel oil)			distilled)
124	.010042	(90% point-gal of diesel fuel oil)			
125	.0010046	(endpoint-gal of diesel fuel oil)			

N.B. At the two TEL levels other than 3.0 ml/gal, the activity levels were identical with those listed above, except as follows:

k_1	1.0 ml/gal	2.9 ml/gal
x_{16}	(not in basis)	0.0043536 (gal of stock 6 in product 1)
x_{21}	0.0036908 (gal of stock 1 in product 2)	.040503 (gal of stock 1 in product 2)
x_{26}	.029526 (gal of stock 6 in product 2)	.025173 (gal of stock 6 in product 2)
x_{101}	.22540 (gal of stock 1)	(not in basis)
(payoff) $\sum_i p_i x_i$	.77142	1.1200

the entire heavy gas oil cut to extinction, but to use only a portion of the light and medium cracked materials for this purpose. The remainder of these two fractions is employed in the blending of three fuel grades — Number 2 heating oil, premium diesel, and Number 6 residual oil.

Seven "disposal activities" enter into the optimal refining program — 111, 118, 119, 120, 123, 124, and 125. It pays the operator to leave some idle cracking capacity in the L coil, and it also pays him to *over*fulfill certain specification requirements on the Number 2 oil product — on the cetane number, the flash point, and the end point. Similarly, with the premium diesel fuel, it is worthwhile to overshoot the quality limits with regard to the specific gravity, the 90 percent point, and the end point of the blend. In the case of all three fuel oils produced — Number 2, diesel, and Number 6 — the viscosity requirement is one that is to be met exactly. Moreover, the cetane number of the diesel blend is a specification that is to be satisfied precisely. A few of these results are intuitively obvious in advance. The remaining ones are at least plausible, once the analysis has been carried through.

Corresponding to the seven "balancing items" in the optimal basis, there are zero shadow prices, u_i, listed for the particular equations. For example, since the program calls for a portion of the L cracking coil to be left idle, the piece of equipment is not a bottleneck, and the imputed *marginal* value of this type of capacity must be zero. By a similar line of reasoning, since the refiner "gives away" cetane numbers along with the Number 2 fuel oil, the imputed cost of meeting this requirement is zero. He does not, however, give away any cetane numbers on the diesel fuel product, and as a reflection of this fact, u_{22} is positive at 0.012647¢ per cetane number-gallon. This means that if the cetane number specification on diesel oil were raised by a "small" amount, say by one cetane number, the change would cost the refiner 0.012647¢ for each gallon of diesel fuel produced. Similarly, the u_i that go along with equations 15, 17, 21, and 26 indicate the incremental cost per gallon of product of altering other particular quality limits.[12]

The variables $u_1, u_2, \ldots, u_{10}$ indicate the marginal values per pound of the intermediate straight-run and cracked oils. Logically enough, the straight-run gasoline (stock 1) is worth less than the higher octane cracked material (stock 6). The other straight-run oils are each valued at more than the cracked material within the same boiling point range. The straight-run

---

[12] Paradoxically, there is a nonzero u_i variable associated with row 16 — the viscosity of Number 1 fuel oil. This end item is one that is not produced at all, and it is hard to give a commonsense interpretation of the positiveness of this Lagrangean multiplier. The particular result is, nevertheless, consistent with condition (3), discussed in this chapter, part 4.1 — the test for optimality of a given basis in a linear programming problem.

streams each have lower viscosities and higher potential gasoline yields than their cracked counterparts.

In many ways, the shadow prices are as interesting as the actual activity levels, x_j. For example, the u_i connected with the various fuel oil streams might be of use in calculating a break-even point on some end item that had not been included specifically within the analysis, e.g., for the Navy Special grade of bunker oil. The imputed values might also be helpful in estimating the worth of additional quantities of a different type of crude than that which was originally postulated for the problem.

The u_i that go along with the equipment capacity limitations — equations 11–14 — may be compared with the marginal cost of expanding the particular facility. For example, u_{13} (the imputed value of coil H capacity) is over twice as great as u_{12} (coil M). At the same time, u_{11} (coil L) is zero. If the refinery were interested in breaking one of these bottlenecks by means of an expansion program, coil H clearly deserves attention. It is true that a new investment problem seldom involves "small" departures from the existing plant. Usually it will be more appropriate to handle such a situation through the parametric linear programming technique discussed below for run number 4.

6. RUN NUMBER 2, VARIATIONS IN THE PRICE OF NUMBER 6 FUEL OIL

Having once calculated an optimal program for the original problem posed by Table VI.5, it now becomes possible to inquire how things might have turned out if some of the assumptions had been different. Suppose, for example, that the price of Number 6 fuel oil (p_{9i}) had come out much higher than the 2.26¢ per gallon figure used originally. It is important to be able to trace through the complex effects of such a change upon the entire manufacturing operation.

Fortunately, it is not necessary to solve out a completely new simplex problem for each alternative value of p_{9i}. Instead, the method of parametric linear programming (discussed in the mathematical appendix to this chapter) makes it easy to determine an optimal schedule corresponding to each such level of the payoff coefficients, p_{9i}. The essential idea is to set the price of

Number 6 fuel oil equal to a constant plus an unknown parameter, θ:

$$p_{9i} = 2.26 + \theta \qquad (i = 0, 2\text{--}5, 7\text{--}9). \qquad (VI.26)$$

The scalar parameter θ is then varied over whatever range looks interesting. The following general principles govern such a situation: (1) if the parameter θ is kept *within* a region for which the initial basis remains an optimal one, then the activity levels, x_j, remain constant. Within these limits of θ, there will be no variation in the output of any of the end products. (2) If the parameter θ crosses *beyond* a certain critical value, there will in general be a new basis which will become optimal. Under this changed basis, a number of the activity levels *may* take on new values. The mere fact that the basis is altered, however, does not require any changes in these input and output levels. Some of the new activities may make their first appearance with a zero value. (3) *At* the critical value of θ, the old and the new basis yield the identical payoff. In fact, at this value, any schedule formed by linear interpolation between the programs of the two bases is itself an optimal one.

To be more concrete, consider the product-mix summary for run number 2, contained in Table VI.10. As the realization on Number 6 fuel oil ranges from 1.84 through 4.84¢ per gallon, the original product-mix generated in run number 1 remains an optimal one. At the price of 4.84¢, a second product-mix becomes optimal, and continues so until at least a fuel oil price of 7.38¢. Over the interval between 4.84 and 7.38¢, there are six distinct bases generated (steps 1–6), but each of these leads to the same product-mix.

Going on the intuitive notion that cracked fuel oil is a low value by-product of the gasoline cracking operation, the reader will not be surprised to find that a *quadrupling* of the price of bunker fuel oil (from 1.84 to 7.38¢) results in only a 10 percent increase in its output (from 19.80 to 21.60 percent of the crude run). As with the naphtha reforming problem in Chapter IV, part 5, the supply of fuel turns out to be inelastic with respect to its own price.

Following out this intuitive argument leads to some curious results. If the fuel oil were literally an unwanted by-product of

Table VI.10. Comparative Product-Mix Summary, Runs Nos. 1 and 2

	Run number 1 (basic case)	Run number 2, steps 1–6 (increase in the price of Number 6 fuel oil)
Lowest applicable price of Number 6 fuel oil, p_{9i} (cents per gallon)	1.84	4.84
Highest applicable price of Number 6 fuel oil, p_{9i} (cents per gallon)	4.84	7.38
Net output (percent by volume of crude run):		
Butanized gasoline		
Premium	4.69	4.06
Regular	49.41	49.41
Total gasoline	54.10	53.47
Distillate oils		
Number 2 fuel oil	9.00	0
Premium diesel	17.40	25.27
Total distillates	26.40	25.27
Number 6 residual fuel oil	19.80	21.60

the manufacture of gasoline, an increase in the price of bunker oil would stimulate the make of cracked gasoline, and correspondingly depress the output of the other refined products — diesel and Number 2 fuel oil. Just the reverse appears to have occurred. The output of premium gasoline declines; that of regular grade gasoline remains constant; the Number 2 grade product vanishes; and the output of diesel oil increases by approximately one-third. The most remarkable effect of the revised price structure is *not* upon the item for which the price has changed, but upon two of the other end products. This provides further evidence on the closeness of the choice between the manufacture of various grades of distillate fuels.

In order to shift from the product-mix for run number 1 to that of run number 2, a whole series of changes in the refining practice had to be performed simultaneously. (The 26 individual activity levels for each step may be found in Table VI.11.) There were decreases in the amounts of each of the three straight-run distillate oils charged to the cracking stills (x_1, x_2, and x_3), and increases in the amount of each cracked distillate (x_5, x_6, and x_7) employed for recycling. The entire straight-run residuum fraction (x_4) was sent to cracking under both of the price structures.

Table VI.11. Activity Levels, Optimal Bases, Runs Nos. 1 and 2

	Run no. 1 (basic case "activity levels" x_j	Run no. 2, steps 1–6, x_j					
		1	2	3	4	5	6
Lowest applicable price of Number 6 fuel oil (cents per gallon) j	1.84	4.84	4.91	5.01	5.16	6.89	7.38
1	0.35546	0.23760	0.23760	0.23760	0.23760	0.23760	—
2	.80403	.15011	.15011	.15011	.15801	.15801	—
3	.32506	.30715	.30715	.30715	.30715	.30715	—
4	1.0000	1.0000	1.0000	1.0000	1.0000	1.0000	—
5	0.17063	0.18706	0.18706	0.18706	0.18706	0.18706	—
6	.16112	.22324	.22324	.22324	.22324	.22324	—
7	.26752	.26972	.26972	.26972	.26972	.26972	—
16	.0060736	.0052588	.0052588	.0052588	.0052588	.0052588	—
21	.040503	.040503	.040503	.040503	.040503	.040503	—
26	.023453	.023453	.023453	.023453	.023453	.023453	—
68	0	0	0	0	0	a	—
73	a	a	a	0	0	0	—
74	.0044252	0	0	0	0	0	—
77	a	a	0	0	a	a	—
78	.0085554	0	0	0	0	0	—
82	.013845	.016376	.016376	.016376	.016376	.016376	—
83	.0026927	.011678	.011678	.011678	.011678	.011678	—
84	.0072314	.0062739	.0062739	.0062739	.0062739	.0062739	—
87	.0011888	.0019962	.0019962	.0019962	.0019962	.0019962	—
90	.024696	.024976	.024976	.024976	.024976	.024976	—
94	a	.0056919	.0056919	.0056919	.0056919	.0056919	—
97	.0037034	.00038855	.00038855	.00038855	.00038855	.00038855	—
111	0.14248	0.14325	0.14325	0.14325	0.14325	0.14325	—
116	a	a	a	a	a	0	—
118	.050152	a	a	a	a	a	—
119	.012981	0	0	a	a	a	—
120	.0041302	0	a	0	0	0	—
123	.00025073	.00041570	.00041570	.00041570	.00041570	.00041570	—
124	.010042	.011108	.011108	.011108	.011108	.011108	—
125	.0010046	.0057131	.0057131	.0057131	.0057131	.0057131	—
$\sum_i p_j x_j$	1.1204[b]	1.1937	1.1958	1.1988	1.2035	1.2572	—
(payoff at lowest applicable price of Number 6 fuel oil)							

a Not in basis.
b Payoff at 2.26¢ (price for Number 6 fuel oil).

Not only the cracking, but also the blending operations are affected by the increase in the value of Number 6 fuel oil. The diesel oil product has the same components as previously (stocks 2, 3, 4, and 7), but it is made up in noticeably different proportions from these four materials. The composition of the blended Number 6 oil product is altered even further. The principal cutter stock for the cracked residuum here is no longer the cracked light distillate stream (x_{97}), but is rather the heavy straight-run distillate (x_{94}). All of these roundabout changes are needed in order to arrive at a preferred operations schedule for values of Number 6 fuel oil lying between 4.84 and 7.38¢ per gallon.

Within this stated range of price variations, nominally there occur five changes of basis, and therefore five different operating programs. In fact, each of these five changes does no more than put one activity into the basis at a zero level, and remove one that had previously been zero. Step 2, for example, introduces activity 77 at a zero level, and then deletes activity 120 (shown at zero for step 1 of run number 2). Similarly, step 3 brings in activity 73 at zero, and knocks out 119, also previously at a zero level.

Steps 1 through 5 of run number 2 do not represent an interesting class of simplex basis changes. They do, however, correspond to a wide range of variation in the price of Number 6 fuel oil, and it is a remarkable thing that the product-mix is unaffected over this entire range. After all, for the payoff coefficient p_{9i} at a level of 7.38¢ per gallon, the price of Number 6 bunker fuel oil comes within half a cent of that of the Number 2 distillate grade. For further increases of the price of Number 6 fuel oil into the distillate range, it can be expected that there would be a large increase in the output of this product. Indeed, for a sufficiently high premium, it would even pay the refiner to divert the entire production of diesel fuel into the Number 6 oil.

7. RUN NUMBER 3, INTER-PRODUCT FLEXIBILITY BETWEEN GASOLINE AND NUMBER 1 FUEL OIL

In view of the possibilities for recycling the various cracked distillate oils through the plant, the refiner has considerable latitude for shifting about his product-mix. At the same time, with a fixed plant, because of the throughput capacity restrictions, there are limits to this process of substitution. For this reason, it seemed interesting to plot out the maximum amount of Number 1 fuel oil production consistent with several alternative requirement levels for cracked gasoline. Economists usually refer to such a curve as a "production function" or a "transformation function" for the two products. In linear programming language, this is also known as an "efficiency frontier" or a "set of efficient points." Koopmans has defined an efficient point as follows: "A possible point . . . in the commodity space (here the two-dimensional gasoline versus Number 1 fuel oil space) is called efficient whenever an increase in one of its coordinates (the net output of one good) can be achieved only at the cost of a decrease in some other coordinate (the net output of another good)." [13]

The technique of parametric linear programming can be readily applied to establish the full set of efficient points when there are only two end items involved. One of the commodities — cracked gasoline, x_{16} — is assigned the arbitrary payoff coefficient of unity. The other product — Number 1 fuel oil — is priced at the level of the parameter θ. That is, the payoff for each of the activities 62, 63, 67, and 68 is set at θ. All of the remaining columns in the matrix, Table VI.5, are given p_j coefficients of zero. This means that no credit is given for products other than gasoline and Number 1 fuel oil, and that none of the usual processing costs are counted.

For a zero value of θ, the problem is simply one of maximizing the output of cracked gasoline. As the parameter θ is increased, the method automatically takes account of all possible price ratios between the gasoline and the fuel oil. For each of these ratios, there will be at least one point within the efficient set that is optimal. Hence, by considering the entire range of relative prices for the two products, it is possible to trace out the complete efficiency frontier.

Just as in the preceding section, we are dealing here with a problem in which a scalar parameter θ enters linearly into the

---

[13] Tjalling C. Koopmans, "Analysis of Production as an Efficient Combination of Activities," Cowles Commission Monograph No. 13, *Activity Analysis of Production and Allocation*, chap. III, p. 60.

payoff coefficients, p_j. The reader is again reminded of the following general principles: (1) If the parameter θ is kept within a region for which the initial basis remains an optimal one, then the activity levels, x_j, remain constant. Within these limits of θ, there will be no variation in the output of either of the end items. (2) If the parameter θ crosses *beyond* a certain critical value, there will in general be a new basis which will become optimal. Under this new basis, a number of the activity levels *may* take on new values. (3) *At* the critical value of θ, the old and the new basis yield the identical payoff. In fact, at this value, any schedule formed by linear interpolation between the programs of the two bases is itself an optimal one.

Table VI.12. Product-Mix Summary, Run No. 3; Substitution Between Cracked Gasoline and Number 1 Fuel Oil

	A		B		C	
Lowest applicable value of θ	0		0.5042		0.6654	
Highest applicable value of θ	0.5042		.6654		∞	
	Gallons per pound of crude	% by volume of crude	Gallons per pound of crude	% by volume of crude	Gallons per pound of crude	% by volume of crude
Net output						
Butane-free cracked gasoline, x_{16}	0.031654	22.02	0.029780	20.71	0.024485	17.03
Number 1 fuel oil						
x_{62}	0		0		.021480	
x_{63}	0		0		0	
x_{67}	.018320		.021682		.007076	
x_{68}	.001934		.002289		.003372	
Total Number 1 fuel oil	.020254	14.09	.023971	16.67	.031928	22.21

N.B. $p_{16} = 1.0$
$p_{62} = p_{63} = p_{67} = p_{68} = \theta$

Table VI.12 provides certain of the numerical results calculated by these methods, and Figure 18 displays the "efficiency frontier" itself. The output of debutanized cracked gasoline is given on the horizontal axis of the figure, while that of Number 1 fuel oil is on the vertical. For values of θ greater than zero and less than 0.5042, the point labeled A is optimal and therefore "efficient." At the critical value of 0.5042, point B is equally profitable. Indeed, all product-mix points on the A-to-B segment are also optimal at this critical price ratio. Moreover, the slope of the segment is 0.5042.

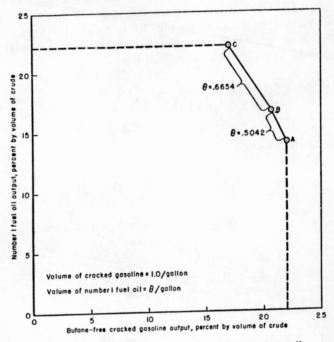

Fig. 18. Run Number 3. Three "Efficient Points"

For values of θ that lie between 0.5042 and 0.6654, again a single product-mix is optimal, that of point B. At the critical level of 0.6654, the payoff remains constant, regardless of whether product-mix B or mix C is employed. Again on this second segment, the slope equals the critical value of θ, a marginal substitution rate of 0.6654 gallons of gasoline per gallon of Number 1 fuel oil. Beyond this second critical value, there is no increase in θ that will alter the output of Number 1 fuel oil. At a maximum, no more than 22.21 percent of the crude can be converted into this product.

It is also worth observing that the *maximum* achievable output of cracked gasoline — 22.02 percent of the crude — is not the level that had been optimal under the representative price structure assumed for run number 1. In the earlier case, only 20.54 percent of the crude was manufactured into the cracked gasoline components, x_{16} and x_{26}. Such results as these should call into doubt any rules of thumb that involve maximizing the output of the highest priced product.

8. RUN NUMBER 4, A NEW INVESTMENT PROBLEM

8.1 INTRODUCTION

All three of the preceding computations — the basic case, the elasticity of supply of Number 6 fuel oil, and the substitution possibilities between gasoline and Number 1 heating distillate — all of these have been grounded upon the assumption that a complete thermal cracking refinery was already in existence, and that there was no possibility for altering the capacities of individual pieces of equipment. The fourth set of calculations, by contrast, is intended to illustrate the problems of planning for an *all-new* refinery. Under such circumstances, it is possible to alter the capacity of the four cracking coils and of the secondary column. As in the three preceding cases, it was taken for granted that a fixed amount of crude (10,000 B/CD) was going to be charged to the primary distillation column. And also as before, the only variety of equipment considered was the thermal cracking type, and not such units as catalytic crackers, reformers, or cokers.

In order to make the analysis comparable with the other three, it was initially stipulated that the over-all investment be the same. After a first solution had been obtained, the capital budget was altered, and the effects upon the optimum program were noted. The outstanding results of this calculation were as follows: Around the initial investment level of $1.8 millions, the marginal return on capital comes to 21 percent per annum, exclusive of depreciation, interest, or maintenance charges. For this investment, by appropriately rearranging the scale of the individual units, the refiner may expect to realize an additional $132,900 in payoff per annum. Or alternatively, for the same payoff as that achieved in the basic case of run number 1, it ought to be possible to reduce the initial investment from $1.80 to around $1.45 millions.

8.2 THE FIXED BUDGET CASE

8.2.1 *The revised linear programming problem.* The first step in performing the "new investment" analysis was to make an estimate of the capital costs associated with each of the five individual pieces of equipment — the four cracking coils and the secondary distillation column. Dr. Singer calculated that the installed cost for the secondary fractionator of the basic case would have amounted to $1,130,000, and that a 3,500 B/CD cracking heater and its accessories would take $220,000. Although these figures constitute the foundation for run 4, the reader would do well to remain skeptical of the particular numbers used here. In transmitting them, Singer wrote as follows:

We have assembled the capital costs which you required to complete the study on maximizing profitability of the simplified refinery. It should be understood that these costs represent only the orders of magnitude that might be expected. The prices that will actually be charged by contractors in any specific case might differ by as much as 50% because of local and current considerations.[14]

For purposes of the cost comparison here, all capital expenses were considered to be directly proportional to the throughput of

[14] Letter dated November 4, 1953.

the individual unit.[15] Taking the cost of a 3,500 B/CD coil at $220,000, one obtains an estimate of $62.86 for each B/CD of cracking charge. This coefficient of $62.86 was in turn applied to the capacities utilized in run number 1, and the results entered in Table VI.13.[16]

Table VI.13. Derivation of Capital Costs for Run No. 4

Equipment item	Pounds of capacity utilized in run no. 1 per pound of crude	Capital cost of equipment item for 10,000 B/CD plant (dollars)	Unit capital cost of capacity (dollars per pound of capacity/CD)
Coil L	0.22252	139,027	0.213948
Coil M	.23750	144,125	.207804
Coil H	.30750	176,207	.196225
Coil R	.37100	206,266	.190384
Secondary column	98.[a]	1,130,000	.0039485[b]
Total capital cost for cracking and recovery facilities		1,795,625	
Capital cost (dollars per pound/CD of crude)		0.614885	

[a] Unit: 10^{-4} moles of vapor flow per pound of crude.
[b] Unit: $/10^{-4}$ moles of vapor flow capacity/CD.

Adding the cost of the four coils to that of the secondary fractionating column, the total comes to $1,795,625 for the cracking section of the 10,000 B/CD refinery, or $0.614885 per pound of crude/CD. In order to maintain comparability between the various calculations, this total excludes any expense for the por-

[15] This assumption of proportionality is the same as that utilized by Adams and Creelman for the naphtha reforming problem, in Chapter IV, sect. 2.2. The economies of scale are considered to be second-order effects.

[16] It can be seen that the capital cost coefficients listed in the table range from $0.190384 up to $0.213948 per pound/CD. The weight-to-volume ratio is different for each of the possible cracking stocks, and despite the constant coefficient of $62.86 per barrel of charge, the cost per *pound* is not invariant. For this part of the calculation, the specific gravity of each of the three straight-run distillates was taken to be identical with that of the cracked material in the same boiling range.

Table VI.14. Changes in Matrix, Table VI.5, for the "New Investment" Problem

Type of activity, x_j Index j	F_R	Balance accounts		Building new capacity					Right-hand side
				Coil L (pounds)	Coil M (pounds)	Coil H (pounds)	Secondary column (10^{-4} moles)	Coil R (pounds)	
	4	127	128	211	212	213	214	227	q_i
Equipment capacity limitations:									
11. Light distillate cracking coil (pounds)				-1					0
12. Medium distillate cracking coil (pounds)					-1				0
13. Heavy distillate cracking coil (pounds)						-1			0
14. Secondary fractionating column (10^{-4} moles)	10.021						-1		0
27. Residuum cracking coil (pounds)	0.371	1						-1	0
Budget restraint:									
28. $ investment/pound of crude/CD	1			0.213948	0.207804	0.196225	0.0039485	0.190384	0.614885

tion of capacity of the light oil cracking unit that had remained idle under the optimal program for run number 1.

Table VI.14 contains the changes used for converting the original matrix, Table VI.5, into a form needed for analyzing the "new investment" problem. Two additional equations were introduced, along with seven activities. In addition, one of the old activities — number 4, F_R — had to be modified. Five of the new activities deal with the construction of new capacity. For example, x_{211} indicates the number of pounds of capacity that are to be built into coil L, and x_{214} the number of 10^{-4} moles of vapor flow capacity per pound of crude oil run. Activities 127 and 128 each constitute balancing items for the two new equations — respectively, 27 and 28.

One of the new rows, number 28, imposes the condition that the total cost of the new construction remain the same as in runs 1 through 3, i.e., within the limit of $0.614885 per pound of crude run. The coefficients on this line are brought over directly from Table VI.13. For example, activity 211 — the building of light oil cracking capacity — requires $0.213948 for each pound of charge/CD. Similarly, activity 214 takes $0.0039485 for each 10^{-4} mole of vapor flow/CD through the secondary distillation column.

The other of the new equations, number 27, ensures that the amount of straight-run residuum scheduled for charging to coil R will not exceed the capacity of that particular unit. In all of the preceding work, since there is no recycle of cracked residuum, it had been possible to guarantee that there would be no bottleneck in the processing of heavy residuum. Under the provisions of run number 4, however, the guarantee is no longer effective. In the construction of a new plant, the refiner is free to divert money away from the residuum cracking coil if he finds more profitable uses for these funds in other parts of the refinery. The coefficient of 0.371 in column 4 and row 27 indicates that if the entire straight-run residuum cut is cracked — i.e., if F_R turns out to be unity — there will be 0.371 pounds of capacity required in coil R per pound of crude run to primary distillation.[17]

Aside from the changes noted in Table VI.14, the "new in-

---

[17] The constant of 10.021 in line 14 and column 4 is identical with that of Table VI.5, but is repeated for the sake of clarity.

vestment" linear programming problem was run off through using a structure identical with the earlier problem displayed in Table VI.5. In retrospect, this did not turn out to be a wise move. Run number 4 could have been performed with five less equations than were actually used. The five equipment capacity limitations, rows 11–14 and 27, are redundant. It would have been possible, for example, to price out the investment required in the light distillate cracking coil and in the secondary fractionating tower per pound/CD of light cycle oil charged. This capital cost could have been associated with activity 5, and the resulting constant entered in the budget restraint row, number 28. The six other cracking and recycling activities could have been handled in a similar fashion. When set up in this more compact way, not only are there five fewer equations, but also six of the new activities are eliminated — numbers 127, 211–214, and 227. In principle, both formulations of the problem are equivalent, but from the viewpoint of computing machine cost, it is regrettable that the alternative statement did not occur earlier.

8.2.2 *An optimum refinery schedule for the fixed budget case.* Table VI.15 presents a summary of the optimum construction program and product-mix for run number 4 — under the condition that the investment budget remain within the ceiling of $1,795,625. For comparative purposes, this program is tabulated alongside the basic results for run number 1. In both cases, the total investment budget was the same, but under the solution of run number 4, the *gross* payoff increased by approximately 1 percent — a differential which would come to $132,900 in the course of a year's operations.

In order to effect this increase in payoff, it was necessary to rearrange all but a few of the refinery's operations. The capacity of the light distillate cracking coil was virtually eliminated, and that of the H coil nearly doubled. It remained optimal to provide enough residuum cracking capacity to accommodate the entire straight-run fraction, and it also remained optimal to provide a capacity of around 28,600 moles of vapor flow/CD for the secondary column.

Although the "new investment" scheme did not entail any substantial changes in the production of gasoline or of Number 6 fuel oil, the output of distillates was altered radically. It became

Table VI.15. Optimum Product-Mix and Construction Program
for the "New Investment" Problem

	Run no. 1 (basic case)	Run no. 4 (new investment problem)
Payoff (cents per pound of crude)	1.12040	1.13287
Net output (% by volume of crude):		
Butanized premium gasoline	4.69	4.52
Butanized regular gasoline	49.41	49.41
Kerosene	0	10.48
Number 1 fuel oil	0	0.86
Number 2 fuel oil	9.00	0
Premium diesel	17.40	15.00
Number 6 fuel oil	19.80	20.35
Charge capacities utilized (10,000 B/CD of crude):		
Coil L (B/CD)	2,212	308
Coil M (B/CD)	2,293	1,844
Coil H (B/CD)	2,803	5,196
Coil R (B/CD)	3,281	3,281
Secondary column (moles of vapor flow/CD)	28,619	28,555

preferable to turn the light straight-run oil directly into kerosene rather than to use it for cracking stock. Conversely, it became worthwhile to increase the use of heavy gas oil cracking stocks at the expense of the Number 2 oil and the diesel fuel end items. Each of these devious shifts had to be planned together when considering the all-new operation.

The reader who is interested in further details on the optimal program should consult Table VI.16. The u_i follow a pattern similar to that of Table VI.9 for run number 1. Perhaps the most interesting of the new "shadow prices" is u_{28}, the imputed marginal return on the capital investment. This amounts to 5.7498% per 100 days, or 20.99 percent per annum.[18] In other words, for

---

[18] This marginal payoff is gross of any incremental charges for maintenance, insurance, taxes, interest, or depreciation.

"small" increases in the over-all investment budget, the annual return to the refinery would increase by some 21 percent.

Table VI.16. Solution to Run No. 4, New Investment Problem

j	"Activity levels," x_i		i	"Shadow prices," u_i	
3	1.0000	F_H	1	1.4722	(¢/pound, SR gasoline)
4	1.0000	F_R	2	1.3065	(¢/pound, SR light oil)
5	0.031015	$R_L Z_L$ (pounds)	3	1.2340	(¢/pound, SR medium oil)
6	.19099	$R_M Z_M$ (pounds)	4	1.2138	(¢/pound, SR heavy oil)
7	.44695	$R_H Z_H$ (pounds)	5	0.71467	(¢/pound, SR residue)
16	.0058563	(gal of stock 6 in product 1)	6	1.7597	(¢/pound, cracked gasoline)
			7	1.2074	(¢/pound, cracked light oil)
21	.040503	(gal of stock 1 in product 2)	8	1.1607	(¢/pound, cracked medium oil)
26	.023453	(gal of stock 6 in product 2)	9	1.1431	(¢/pound, cracked heavy oil)
			10	0.15647	(¢/pound, cracked residue)
52	.015076	(gal of stock 2 in product 5)	11	.012302	(¢/pound capacity, coil L)
67	.0011220	(gal of stock 7 in product 6)	12	.011948	(¢/pound capacity, coil M)
68	.00011844	(gal of stock 8 in product 6)	13	.011282	(¢/pound capacity, coil H)
			14	.00022703	(¢/10⁻⁴ moles capacity, secondary column)
77	0	(gal of stock 7 in product 7)			
78	0	(gal of stock 8 in product 7)	15	.75275	(¢/gal, regular gasoline)
			16	.42713	(¢/μ-gal, Number 1 fuel oil)
82	.0064045	(gal of stock 2 in product 8)	17	3.2954	(¢/μ-gal, Number 2 fuel oil)
83	.013740	(gal of stock 3 in product 8)	18	0	(¢/CN-gal, Number 2 fuel oil)
87	.0014239	(gal of stock 7 in product 8)	19	0.28539	(¢/gal, Number 2 fuel oil)
90	.025457	(gal of stock 10 in product 9)	20	0	(¢/gal, Number 2 fuel oil)
97	.0038176	(gal of stock 7 in product 9)	21	.2518	(¢/μ-gal, diesel fuel oil)
			22	.023618	(¢/CN-gal, diesel fuel oil)
118	0	(CN-gal of Number 2 fuel oil)	23	0	(¢/S.G.-gal, diesel fuel oil)
120	0	(endpoint-gal of Number 2 fuel oil)	24	0	(¢/gal, diesel fuel oil)
			25	0	(¢/gal, diesel fuel oil)
			26	6.8536	(¢/μ-gal, Number 6 fuel oil)
123	.00029145	(S.G.-gal of diesel fuel oil)			
124	.0035156	(90% point-gal of diesel fuel oil)	27	.010947	(¢/pound capacity, coil R)
125	.0071207	(endpoint-gal of diesel fuel oil)	28	5.7498	(% return on capital per 100 days)
211	.031015	(pounds capacity, coil L)			
212	.19099	(pounds capacity, coil M)	(payoff) $= \Sigma\, p_i x_i = 1.13287$ (¢ per pound of		
213	.56995	(pounds capacity, coil H)	crude distilled)		
214	97.781	(10⁻⁴ moles capacity, secondary column)			
227	.37100	(pounds capacity, coil R)			

8.3 PROGRAMMING FOR ALTERNATIVE BUDGET LEVELS

Having determined an optimum construction program for the $1.8 million fixed budget, we are now in a position to calculate the effects of "large" changes in this capital sum. As before, the only refining processes considered are of the thermal cracking type. The product realizations, the yield structure, and all other engineering assumptions remain the same. The only alteration consists of taking the investment budget, not as a constant, but rather as dependent upon an unknown parameter θ:

$$q_{28} = 0.614885 + \theta. \qquad (VI.27)$$

In other words, by varying θ over a positive and a negative range, it is possible to trace out the implications of an altered budget for the entire product-mix and capacity construction problem. In arriving at an optimal program for a changed value of θ, the scheduler does not repeat the entire simplex routine from scratch. Instead, as shown in the mathematical appendix following this chapter, he may calculate the effects of a new θ with little more effort than is required for a single shift in basis.

There are three general principles governing this type of parametric linear programming, and they are analogous to those already stated for the case in which the parameter entered linearly into the payoff coefficients. Now, for the "dual" problem, in which θ goes linearly into the right-hand-side coefficients, q_i, these principles are as follows: (1) If θ is kept *within* a region for which the initial basis remains a feasible one, then the shadow prices, u_i, remain constant. For constant u_i, the payoff will not be improved by bringing into the basis any activity that had previously lain outside. (2) If the parameter θ crosses *beyond* a certain critical value, there will in general be a new basis which will become both feasible and optimal. (3) *Between* two neighboring critical values of θ, any schedule formed by linear interpolation will be optimal for the corresponding intermediate level of θ.

When considered in the context of the "new investment" problem, these abstract-sounding principles take on more life. (See Table VI.17.) As θ moves between the critical values of -0.029772 and $+0.025482$ — that is, so long as the total investment remains between \$1,708,682 and \$1,870,038 — the simplex basis remains the same. The marginal return on capital continues to be 20.99 percent per annum, and all the other shadow prices are constant. Between these two critical values (columns B and D), the activity levels themselves change in a linear fashion or else remain fixed. The output of premium gasoline, for example, moves up linearly from 3.74 percent of the crude run to 5.20 percent. The output of regular grade gasoline stays at a constant level; that of Number 1 fuel oil declines steadily, and there are compensating alterations made in the designed capacities for the various pieces of equipment.

As θ increases beyond the critical level of \$0.025482 per pound of crude run, in order to preserve "feasibility" it is necessary to introduce into the program a hitherto unused activity, F_M, the cracking of straight-run medium distillate oil. By the same token, activity 67 is discarded.

Within the new basis, for values of θ between 0.025482 and 0.041869, the u_i shadow prices again remain constant, and the x_j activity levels move linearly. Throughout this range, the marginal return on capital is 20.46 percent; some output of diesel fuel is sacrificed, and the make of premium gasoline is increased still further.

Table VI.17. Product-Mix, Charge Capacities, and Payoff at Five Levels of Capital Investment

	A	B	C	D	E
Total investment for 10⁴ B/CD of crude	1,637,685	1,708,682	1,795,625	1,870,038	1,917,892
θ (dollars per pound of crude)	−0.054084	−0.029772	0	0.025482	0.041869
Payoff (cents per pound of crude)	1.12932	1.13116	1.13287	1.13434	1.13525
Marginal return on capital (percent per annum)	27.64		20.99		20.46
Net output (percent by volume of crude):					
Butanized premium gasoline	3.07	3.74	4.52	5.20	5.68
Butanized regular gasoline	49.41	49.41	49.41	49.41	49.41
Kerosene	9.07	10.48	10.48	10.48	10.74
Number 1 fuel oil	2.89	1.87	0.86	0	0
Number 2 fuel oil	0	0	0	0	0
Premium diesel	15.42	15.00	15.00	15.00	14.12
Number 6 fuel oil	20.08	20.23	20.35	20.46	20.54
Charge capacities (pounds per pound of crude)					
Coil L	0	0	0.031015	0.05750	0.06105
Coil M	0.14992	0.18313	.19099	.19772	.21692
Coil H	.56110	.56575	.56995	.57355	.57604
Coil R	.37100	.37100	.37100	.37100	.37100
Secondary column (10⁻⁴ moles)	88.366	92.544	97.781	102.26	105.09

Along with each of the four critical values of θ listed in Table VI.17 (columns A, B, D, and E), there is now associated a particular level of gross payoff. With these four points, it is possible to establish a payoff-versus-investment curve over an interval of some $280,000 in investment cost. (See Figure 19.) For all investment levels displayed on this chart, the gross profit corresponding to the "new investment" problem, run number 4, exceeds that of the basic case, run number 1. Indeed, even employing a pessimistic extrapolation, the level of total investment could be trimmed to $1.45 millions before the payoff was reduced to the same magnitude as that of run number 1. *If* then these yield coefficients and capital cost estimates are correct ones, the refiner might have saved $350,000 in his initial construction costs, and yet realized the same gross income as in the original case where the unit capacities were assigned by rules of thumb.

A capital saving of this magnitude might well be a freak occurrence, but there is certainly no reason for supposing that a 1:1.5 recycle ratio principle is superior to a linear programming analysis. The formal approach not only reveals an operating program with a higher payoff to the refiner. It also yields up its results in a form that would be directly useful to the financial officers of a company. Instead of confronting these officials with an either-or choice of processing schemes such as in Table II.1, a refinery engineering staff could easily derive a whole series of alternative programs, each of them optimal for a different level of capital investment. In this way, those at a higher echelon in the company would find it easier to arrive at the major financial policy decisions — for example, the choice between an expansion of refinery facilities and the repayment of short-term indebtedness. Top management is continually perplexed by such matters as these. A mathematical analysis of the refinery operations problems would make it possible for the engineer to present his findings in an easily digestible form. At the same time, this framework enables the scheduler to consider the interlocking effects

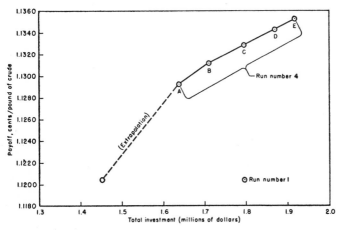

Fig. 19. Payoff-Investment Curve, Run Number 4

of the many variables that are connected with his decisions. Neither nomogram methods nor educated guesses are likely to remain satisfactory for dealing with an integrated refinery problem that involves upwards of sixty independent unknowns.

9. SUMMARY AND CONCLUSIONS

This chapter has been concerned with a simplified version of a hypothetical representative thermal cracking refinery located in Oklahoma. It was possible to convert into the language of linear programming an engineering analysis that had originally involved numerous nonlinearities. Once the linear programming system was put together, four sets of calculations were performed. The first of these took for granted both an existing plant and an existing market price structure. The straightforward solution yielded an optimal refinery schedule. As a general check upon the original engineering assumptions, it was reassuring to find that the preferred output of gasoline and of distillate fuel resembled that of the actual Oklahoma refining region.

The second calculation was designed to show the effects of a rise in the price of the Number 6 fuel oil end product. Within the price range considered, it turned out that a quadrupling of the realization on this item would lead to only a 10 percent increase in the optimal amount to be supplied, but that there would be significant side-effects upon the output of two of the distillate oil products.

The third computing run was addressed to the question of production flexibility between two of the chief end items — gasoline and Number 1 fuel oil. This calculation indicated that, for a *given* plant, there are definite limits to the possibilities of converting the one product into the other. Regardless of the relative demand for the two, this refinery could not turn more than 22.02 percent of its crude oil into cracked gasoline, nor more than 22.21 percent into Number 1 fuel oil. Within the stated range, approximately sixty barrels of gasoline yield may be obtained at the expense of one hundred barrels of the distillate fuel.

The final computation dealt with the problem of constructing an all-new thermal cracking refinery. For the same over-all initial investment as that which had been previously assumed — $1.8

millions — it was possible to redesign the unit capacities in such a way that the refinery might have earned an additional $132,900 per annum. Or, for the same level of gross return, it would have been possible to cut the initial investment by $350,000. Not only would the mathematical programming approach have made it possible to improve the expected profitability of the plant; it also would have provided top management with knowledge of the payoff implications for various sizes of the capital budget.

Again it should be emphasized that the mathematical analysis does not eliminate the fundamental uncertainties facing an oil company — the future demand for its products, the behavior of its rivals, the future investment opportunities, and the attitude of public regulatory bodies. Rather, the scheduling calculation enables the refinery people to present the executive group with a detailed program for dealing with each of several possible contingencies. With such information readily at hand, the area of uncertainty on the major problems ought at least be reduced.

Appendix to Chapter VI.

NOTES ON PARAMETRIC LINEAR PROGRAMMING[20]

I. *The general linear programming problem*

The general linear programming problem may be stated as follows:

$$\text{Maximize } \sum_{j=1}^{n} p_j x_j,$$

$$\text{subject to: } \sum_{j=1}^{n} a_{ij} x_j = q_i \quad (i = 1, 2, \ldots, m)$$

$$x_j \geq 0 \quad (j = 1, 2, \ldots, n). \quad \text{(VI.28)}$$

In the purely static cracking-recycling-blending problem, $m = 26$, and $n = 61$. In the "new investment" problem, these become 28 and 68, respectively.

Using matrix notation, the system (VI.28) becomes:

$$\text{Maximize } P' X$$

$$\text{subject to: } AX = Q$$

$$X \geq 0. \quad \text{(VI.29)}$$

The column vectors P, X, and 0 have n components. The column vector Q contains m elements. The matrix A includes n columns and m rows. The column vectors within A are labelled, respectively, $A_1, A_2, \ldots, A_n$.

II. $P(\theta)$

In the case of the price variation and the "efficiency frontier" problems, the payoff vector depends linearly upon the *scalar* θ:

$$p_j(\theta) = p_j + c_j \theta \quad (j = 1, 2, \ldots, n). \quad \text{(VI.30)}$$

---

[20] These notes are based upon hitherto unpublished work of George Dantzig, Alan Hoffman, Carlton Lemke, Harry Markowitz, and William Orchard-Hays.

Or, in matrix form,

$$P(\theta) = P + C\theta. \qquad (VI.31)$$

The optimizing problem becomes:

$$\text{Maximize} \quad P(\theta)' X,$$
$$\text{subject to:} \quad A X = Q$$
$$X \geq 0. \qquad (VI.32)$$

Suppose we have determined a "basic feasible" vector X^0, associated with a particular basis B. (B is an $m \times m$ nonsingular matrix, whose columns represent a subset of A.) From (VI.29), we have:

$$B X^0 = Q$$
$$X^0 \geq 0. \qquad (VI.33)$$

Furthermore, there exists a dual vector, $U^0(\theta)$, defined as follows:

$$B' U^0(\theta) = P + C\theta. \qquad (VI.34)$$

(The vectors P and C are formed from the p_j and c_j associated with those A_j included within the basis, B.)

Solving (VI.34) for $U^0(\theta)$:

$$U^0(\theta) = (B')^{-1}(P + C\theta). \qquad (VI.35)$$

Suppose now that X^0 constitutes an "optimal" vector for $\theta = 0$. Then for any variation of θ such that the basis B remains optimal,

$$A' U^0(\theta) \geq P(\theta) \qquad \text{(all } A_j\text{)}, \qquad (VI.36)$$

or $\quad A' U^0(\theta) \geq P + C\theta \qquad \text{(all } A_j\text{)}, \qquad (VI.37)$

or $\quad A_j' U^0(\theta) \geq p_j + c_j\theta \qquad (j = 1, 2, \ldots, n). \qquad (VI.38)$

Substituting the expression in (VI.35):

$$A_j'(B')^{-1}(P + C\theta) \geq p_j + c_j\theta \qquad (j = 1, 2, \ldots, n). \qquad (VI.39)$$

Define the m-component vector D_j as follows:

$$D_j = A_j'(B')^{-1}. \qquad (VI.40)$$

Combining (VI.39) and (VI.40), at a zero value of θ:

$$D_j P \geq p_j \qquad (j = 1, 2, \ldots, n). \qquad (VI.41)$$

From (VI.33), it can be seen that the vector X^0 remains feasible, regardless of variations in θ. As θ is altered, however, the equation (VI.34) indicates that the basis need *not* remain "optimal." At the following level of θ, we will be exactly indifferent whether or not to introduce one of the previously excluded vectors, A_j:

$$A'_j U^0 (\theta) = p_j + c_j \theta \qquad (A_j \sim \epsilon B).[19] \qquad \text{(VI.42)}$$

Substituting as in (VI.39)

$$D_j P + D_j C \theta = p_j + c_j \theta \qquad (A_j \sim \epsilon B). \qquad \text{(VI.43)}$$

$$\theta = \frac{p_j - D_j P}{D_j C - c_j} \qquad (A_j \sim \epsilon B). \qquad \text{(VI.44)}$$

Call Δ the maximum increase in θ which can take place without introducing one of the previously excluded vectors. From (VI.41), it is known that $(p_j - D_j P)$ is nonpositive. If Δ is to be positive, the only vectors to be considered are those with negative $(D_j C - c_j)$.

The criterion for choosing Δ is as follows:

$$\Delta = \underset{A_j \sim \epsilon B}{\text{minimum}} \left\{ \frac{p_j - D_j P}{D_j C - c_j} ; (D_j C - c_j) < 0 \right\}. \qquad \text{(VI.45)}$$

Once Δ is determined, the algorithm also indicates A_j, the vector to be introduced into the basis. From this point on, the usual simplex routine is employed. The new set of X^0 and $U^0 (\theta)$ is calculated, and we again determine the maximum increase in θ consistent with optimality for the new basis. In case there are any "ties" in choosing A_j through the criterion (VI.45), the vector is chosen by an arbitrary convention.

III. $Q (\theta)$

In the case of the "new investment" problem, the *right-hand-side* vector depends linearly upon the scalar θ:

$$q_i (\theta) = q_i + k_i \theta \qquad (i = 1, 2, \ldots, m). \qquad \text{(VI.46)}$$

---

[19] The expression $(A_j \sim \epsilon B)$ is to be interpreted as "the vector A_j is *not* included in the basis B." Using similar notation, the expression $(A_j \epsilon B)$ means the vector A_j *is* included in the basis B.

In matrix notation:

$$Q(\theta) = Q + K\theta. \tag{VI.47}$$

This leads to the following optimizing problem:

Maximize $P'X$,

subject to $AX = Q(\theta)$

$$X \geq 0. \tag{VI.48}$$

Suppose that an optimal, feasible solution, X^0, has been found for a zero value of θ. The vector X^0 is associated, as before, with a nonsingular $m \times m$ matrix B:

$$BX^0 = Q(0). \tag{VI.49}$$

Define the m-component vector Y as follows:

$$BY = -K, \tag{VI.50}$$

or $$Y = B^{-1}(-K). \tag{VI.51}$$

As θ is increased, the matrix B remains "optimal," but need not stay feasible. One of the previously positive x_j^0 will be driven below zero. Then A_r, the vector to be removed from the basis, may be determined according to the usual simplex criterion. In addition, this yields θ_r, the maximum increase in θ consistent with keeping A_r within the basis.

$$\theta_r = \frac{x_r^0}{y_r} = \operatorname*{minimum}_{j \,\epsilon\, B} \left\{ \frac{x_j^0}{y_j} ; y_j > 0 \right\}. \tag{VI.52}$$

(N.B. If all $y_j \leq 0$, then θ can be made arbitrarily large, and the existing basis will remain feasible as well as optimal.)

Adding and subtracting $(\theta_r B Y)$, the equation (VI.49) now reads:

$$B(X^0 - \theta_r Y) + \theta_r(-K) = Q. \tag{VI.53}$$

All components of $(X^0 - \theta_r Y)$ are nonnegative, and the rth one equals zero.

Adding $\theta_r K$ to both sides of (VI.53)

$$B(X^0 - \theta_r Y) = Q + \theta_r K. \tag{VI.54}$$

The vector $(X^0 - \theta_r Y)$ is still feasible and optimal for $Q(\theta_r)$.
It is now necessary to select the vector A_k that must be introduced into the basis in order to preserve feasibility for $\theta > \theta_r$.
Since the previous basis had been an optimal one:

$$A'_j U^0 \geq p_j \qquad (j = 1, 2, \ldots, n). \qquad (\text{VI.55})$$

Suppose that A_k, a previously excluded vector, is to be introduced into the basis. This vector's representation in the old basis is an m-component column vector termed Y_k:

$$Y_k = B^{-1} A_k. \qquad (\text{VI.56})$$

If the vector β_r is the rth row of B^{-1},

$$y_{rk} = \beta_r A_k. \qquad (\text{VI.57})$$

If A_k replaces A_r in the basis, there will come about a new set of U^0. These are termed U^1. Again, in order to preserve optimality:

$$A'_j U^1 - p_j \geq 0 \qquad (j = 1, 2, \ldots, n). \qquad (\text{VI.58})$$

By definition of U^1, (VI.58) holds with an equality sign for all A_j vectors in the new basis.

Next, it is observed that by definition of U^0:

$$U^{0'} A_j - p_j = 0 \qquad (\text{all } A_j \epsilon B). \qquad (\text{VI.59})$$

And also, by definition of the inverse matrix:

$$\beta_r A_j = 0 \qquad (\text{all } A_j \epsilon B, \text{ but not } A_r). \qquad (\text{VI.60})$$

From (VI.59) and (VI.60), the following is true:

$$\left[U^{0'} - \left(\frac{U^{0'} A_k}{\beta_r A_k} \right) \beta_r + \frac{p_k}{\beta_r A_k} \beta_r \right] A_j - p_j = 0 \qquad (\text{VI.61})$$

$$(\text{all } A_j \epsilon B, \text{ but not } A_r; \text{ also } A_j = A_k).$$

The vector in parentheses preceding A_j must be identical with U^1. It satisfies the definition of U^1, and the latter is unique.
Rewriting (VI.58) with the new expression for U^1:

$$(A'_j U^0 - p_j) - \frac{\beta_r A_j}{\beta_r A_k} (A'_k U^0 - p_k) \geq 0$$

$$(j = 1, 2, \ldots, n). \qquad (\text{VI.62})$$

Now consider only the negative $\beta_r A_j$. Since $(A_j' U^0 - p_j) \geq 0$:

$$\frac{A_j' U^0 - p_j}{\beta_r A_j} \leq 0. \qquad (VI.63)$$

Dividing (VI.62) by $\beta_r A_j$, a negative quantity, this implies:

$$\frac{A_j' U^0 - p_j}{\beta_r A_j} \leq \frac{A_k' U^0 - p_k}{\beta_r A_k} \leq 0. \qquad (VI.64)$$

Or,

$$\frac{A_k' U^0 - p_k}{-\beta_r A_k} = \underset{j \,\epsilon\, B}{\text{minimum}} \left\{ \frac{A_j' U^0 - p_j}{-\beta_r A_j}; \beta_r A_j < 0 \right\}. \qquad (VI.65)$$

If all $\beta_r A_j \geq 0$, it can be shown that no feasible solution exists for $\theta > \theta_r$.

The last step is to introduce A_k and to delete A_r. If the index r is unique, then with the new basis the new θ can be made to exceed θ_r.

Chapter VII

The Economist and the
Operations Scheduler

1. What can the economist currently contribute to the refinery programming problem?

The four case studies presented in this volume have been intended to illustrate some of the potentialities, and pitfalls, in applying mathematical economics to the operation of actual business activities. The first three of these studies, the allocation of crude oils, the operation of a naphtha conversion unit, and the blending of gasolines, are each focused upon narrow segments of a larger problem. This larger problem consists of coördinating the operations of an integrated oil company — from crude petroleum production through transportation and refining to marketing. As a step in the direction of planning for an integrated concern, the fourth study brings together several of the conversion and the blending stages of the manufacturing process.

In progressing beyond this point, in principle it ought not be difficult to include additional types of crude oils and of conversion units — catalytic cracking, alkylation, lube oil processing, and so on. These further steps might involve the gathering of additional laboratory information, but would seldom require radically new data. Most of the necessary numbers are already in the files of the major refiners and of the process engineering companies.

Once this basic technological work has been done, it should be profitable to extend the programming methods so as to take more adequate account of geography and of time. This volume has generally been concerned with the mechanics of a steady-state process, and has also ignored spatial details, e.g., crude oil pipe-line capacity limitations.

In connection with many problems, these simplifications may be legitimate. However, in determining an optimal production policy for an item such as domestic furnace oil, the refiner must consider the possibility of carrying inventories forward from the slack summer season to the active winter season. He cannot afford to ignore the interaction between the winter and the summer operating program.

Geography enters the scheduling problem in a similar fashion. Whenever two plants interchange blending stocks, it is preferable to recognize explicitly any transport costs of the operation, and not follow the shortcut procedure adopted in Chapter V for handling the Oleum-Wilmington exchange of raw gasolines.

Numerous mathematical devices are available for handling refinery scheduling problems — graphical methods, the calculus, linear programming, and several species of nonlinear program-ming. For many of the applications, it is possible to obtain a solution rapidly and cheaply by hand methods. (Witness the crude oil allocation and the naphtha reforming problems.) In other instances, the electronic calculator has a commanding ad-vantage, and provides the only feasible means of accomplishing the mathematical routine.

2. What is the economist currently unable to contribute to the refinery programming problem?

However attractive may be the newer tools of numerical analy-sis, the would-be scheduler is plagued by the traditional diffi-culties of economics work. He is not in a good position to forecast either the demand or the prices for refined products. He cannot even be sure of the future quality specifications on these items. And since he does not know what investment opportunities will occur in the future, the scheduler is handicapped in making any decisions on the magnitude of investment outlays.

These marketing and finance problems do not fit neatly into the currently known body of scientific knowledge, and the most sensible course appears to consist of the present policy, referring them to top-level management. Despite this generally negative conclusion, the mathematical programmer is in a good position to supply information relevant to these decisions, and needed by those in the higher echelons. He should, for example, be able to determine the company-wide cost implications of an octane number increase, and he might even be able to recommend how to split up an investment budget among competing projects for alternative levels of that budget. Information of this nature is difficult to obtain through any of the conventional methods of cost analysis, but it does not impose extraordinary burdens upon mathematical programming.

3. What can the economist learn about the refinery operator?

To the general economist, perhaps the most interesting thing that emerges out of the current study is a reminder that the classical problems of resource allocation still remain with him. The petroleum industry provides a fascinating array of cases in which there are alternative methods of production for achieving "given ends." The refiner must choose among scores of possible ways of controlling his equipment, his raw materials, and his products.

The engineer and the mathematical economist, by working together, may succeed in narrowing down the area of choice among alternatives. For the purely static problems, there are no conceptual difficulties involved in tracing out the set of efficient points. This does not mean that the economist is even close to the stage at which he can perform a mechanical prediction of the behavior of an independent business entity. In order to approach this point, it will be necessary for him to acquire a far more intimate understanding of how the "given ends" are themselves determined.

4. How can the economist make better estimates of sector-wide and economy-wide capabilities?

Despite all difficulties in predicting the behavior of decentralized units operating within a market framework, the outlook for the economist is far from gloomy. As he acquires a more intimate understanding of engineering processes, and as he acquires more skill in applying the techniques of numerical analysis, he can afford to place more confidence in his forecasts of production capabilities — whether of individual sectors or of an entire economy. This does not mean that the economist is in any position to draw up a detailed master scheme for the operation of an economy such as the United States. Rather, it means that he should be able to predict whether a given program of output for consumption, investment, and national defense is consistent with the existing resource limitations and the engineering state of the art. It is my guess that feasibility-testing, rather than the detailed forecast of market behavior, will continue for a long time to be the chief reason for constructing economy-wide models of production optimization.

Work is now underway on "process analysis" studies for a number of individual sectors: metal-working, iron and steel, electric power, chemicals, and petroleum.[1] It is expected that these intra-sectoral analyses will not only be useful for feasibility-testing of a limited scope, but also that they can be joined together as elements of an economy-wide model. In each of the intra-sectoral studies, the linear programming matrix is being based largely upon engineering-type data, rather than upon time series. Through this approach to the problem of resource substitution, it should be possible to overcome many of the inconsistencies that have occurred in previous attempts to build interindustry models.

[1] For a general description of this project, see Harry Markowitz, *The Nature and Applications of Process Analysis*, RM-1254, The RAND Corporation, Santa Monica, 24 May 1954.

Index

Absolute advantage, principle of, 34–36, 41–42
Adams, N. R., 45–69, 161n
Algorithm, definition of, 5
Alkylate, definition of, 5
Antiknock agents, definition of, 5
API gravity, definition of, 5
Arrow, K., 110, 125
ASTM, definition of, 6
ASTM distillation, definition of, 6

Bain, J. S., 71n
Barrel, definition of, 6
Basis in a linear programming problem, 32–43, 127, 135, 152–177
Beckman, M., 82
Bogen, J. S., 77–78
Bottoms, definition of, 6
Btu, definition of, 6
By-product method of cost allocation, 11–14

Catalyst, definition of, 6
Catalytic cracking, definition of, 6
Centistoke, definition of, 6
Cetane number, definition of, 6
Charnes, A., 21n, 32, 79n, 127n
Chipman, J., 128n
Clark, J. M., 102n
Coking, definition of, 6
Comparative advantage, principle of, 41, 43–44

Cooper, W. W., 21n, 79n, 127n
Cracked gas oil, definition of, 6
Cracking, definition of, 6
Creelman, G. D., 45–69, 161n

Dantzig, G. B., 21, 32, 38, 44, 90, 127–128, 146, 172
Degeneracy in linear programming problems, 31–32, 34n
Distillation, definition of, 6
Dorfman, R., 28n, 110n, 127n
Dual variables in a linear programming problem, 33–34, 38–43, 167

Eastman, D. B., 77
Economies of scale, 48–50, 161n
Efficient points, 90, 97, 156–159, 180
Elasticity of supply (demand), 6, 57–60, 152
Empty vessel problem, 22, 27
End point (EP), definition of, 7
Esso Standard Oil Company, 21

Fractional distillation, definition of, 7

Garton, F. L., 76, 82n
Gary, W. W., 45n
Gas oil, definition of, 7
Gasoline, definition of, 7
Gasoline replacement value formula, 14–16
Gasoline specifications, 72–73

Hebl, L. E., 76, 82n
Henderson, A., 127n
Hitch, C. J., 106n
Hitchcock, F. L., 21n, 22, 38, 44
Hoffman, A., 172n
Hornaday, G. F., 16n
Houdry Process Corporation, 16

Identity matrix, definition of, 7
Initial boiling point (IBP), definition
    of, 7
International Business Machines Card
    Programmed Calculator, 73–74, 82,
    91, 128, 146
International Business Machines Cor-
    poration, "701" Data Processing
    System, 74n

Joint costs, 10, 26

Kerosene, definition of, 7
Kinematic viscosity, definition of, 7
Koopmans, T. C., 21n, 22n, 38, 44, 90,
    127, 156
Kuhn, H. W., 81–83, 89–90, 95

Lagrangean multiplier, 7, 81n, 85, 94–
    96, 104–106, 150n
Lange, O., 83, 106n
Lemke, C., 172n
Lerner, A. P., 83, 102, 106n

McCreery, A. R., 71n
McKee, R. W., 11
Manne, A. S., 78n
Market limitations on sales, 47, 71–72,
    98–103, 124, 148
Markowitz, H., 82, 172n, 181
Marshall, A. W., 106n
Matrix, definition of, 7
Mellon, B., 21n, 79n
Middle distillates, definition of, 7
M. W. Kellogg Company, 45, 79, 105

Naphthas, definition of, 7
Nelson, W. L., 124
Neyman, J., 81n
Nichols, R. M., 77–78
Noll, N. D., 16n
Norton, J., 70n

Octane number, definition of, 8
Oklahoma, Kansas, Missouri refining
    region, 146–147
Oleum, California, 73, 179
Opportunity cost, definition of, 8, 14–
    15, 102
Orchard-Hays, W., 127–128, 146, 172
Orden, A., 32

Parameter, definition of, 8
Payoff, definition of, 8
Peavy, C. C., 16n
Polymerization, definition of, 8

Reaugh, M., 70n
Recycling, definition of, 8
Reduced crude, definition of, 8
Reforming, definition of, 8
Reid Vapor Pressure, definition of, 8
Reiter, S., 22n
Relative crude evaluation method, 23–
    26
Rendel, T. B., 76, 82n
Residual fuel oils, definition of, 8
Residuum, definition of, 8

Sachanen, A. N., 125
Sales value method of cost allocation,
    11–14
Samuelson, P. A., 28n, 67n
Scalar, definition of, 8
Shadow price, 8, 33, 81n, 83, 102, 104–
    106, 127, 148–151, 162–167
Shell Development Company, 109, 110
Shell Oil Company, 110n
Singer, E., 109, 110, 124, 125, 137, 146,
    160
Socony Vacuum Oil Company, 13
Solow, R., 28n
Standard Oil Company (Indiana), 4, 14
Standard Oil Company (New Jersey),
    14
Straight-run gasoline, definition of, 9

Temporary National Economic Com-
    mittee, 13, 14
Tetraethyl lead, definition of, 9
Texas Corporation, 13
Thermal cracking, definition of, 9
Tucker, A. W., 81–83, 89–90, 95

Union Oil Company of California, 70–108, 135n, 140
Univac Fac-Tronic System, 74n
U. S. Bureau of Mines, 147
U. S. Bureau of Naval Personnel, 5n
U. S. Office of Price Administration, 13n

Vector, definition of, 9
Visbreaking, definition of, 9
Viscosity, definition of, 9

Weinrich, W., 16n
Wilmington, California, 72, 179
Wilson, R. E., 4, 14